Springer Complexity

Springer Complexity is an interdisciplinary program publishing the best research and academic-level teaching on both fundamental and applied aspects of complex systems – cutting across all traditional disciplines of the natural and life sciences, engineering, economics, medicine, neuroscience, social and computer science.

Complex Systems are systems that comprise many interacting parts with the ability to generate a new quality of macroscopic collective behavior the manifestations of which are the spontaneous formation of distinctive temporal, spatial or functional structures. Models of such systems can be successfully mapped onto quite diverse "real-life" situations like the climate, the coherent emission of light from lasers, chemical reaction-diffusion systems, biological cellular networks, the dynamics of stock markets and of the internet, earthquake statistics and prediction, freeway traffic, the human brain, or the formation of opinions in social systems, to name just some of the popular applications.

Although their scope and methodologies overlap somewhat, one can distinguish the following main concepts and tools: self-organization, nonlinear dynamics, synergetics, turbulence, dynamical systems, catastrophes, instabilities, stochastic processes, chaos, graphs and networks, cellular automata, adaptive systems, genetic algorithms and computational intelligence.

The two major book publication platforms of the Springer Complexity program are the monograph series "Understanding Complex Systems" focusing on the various applications of complexity, and the "Springer Series in Synergetics", which is devoted to the quantitative theoretical and methodological foundations. In addition to the books in these two core series, the program also incorporates individual titles ranging from textbooks to major reference works.

Understanding Complex Systems

Founding Editor: J.A. Scott Kelso

Future scientific and technological developments in many fields will necessarily depend upon coming to grips with complex systems. Such systems are complex in both their composition – typically many different kinds of components interacting simultaneously and nonlinearly with each other and their environments on multiple levels – and in the rich diversity of behavior of which they are capable.

The Springer Series in Understanding Complex Systems series (UCS) promotes new strategies and paradigms for understanding and realizing applications of complex systems research in a wide variety of fields and endeavors. UCS is explicitly transdisciplinary. It has three main goals: First, to elaborate the concepts, methods and tools of complex systems at all levels of description and in all scientific fields, especially newly emerging areas within the life, social, behavioral, economic, neuro- and cognitive sciences (and derivatives thereof); second, to encourage novel applications of these ideas in various fields of engineering and computation such as robotics, nano-technology and informatics; third, to provide a single forum within which commonalities and differences in the workings of complex systems may be discerned, hence leading to deeper insight and understanding. UCS will publish monographs, lecture notes and selected edited contributions aimed at communicating new findings to a large multidisciplinary audience.

New England Complex Systems Institute Book Series

Series Editor

Dan Braha

New England Complex Systems Institute
24 Mt. Auburn St.
Cambridge, MA 02138, USA

New England Complex Systems Institute Book Series

The world around is full of the wonderful interplay of relationships and emergent behaviors. The beautiful and mysterious way that atoms form biological and social systems inspires us to new efforts in science. As our society becomes more concerned with how people are connected to each other than how they work independently, so science has become interested in the nature of relationships and relatedness. Through relationships elements act together to become systems, and systems achieve function and purpose. The study of complex systems is remarkable in the closeness of basic ideas and practical implications. Advances in our understanding of complex systems give new opportunities for insight in science and improvement of society. This is manifest in the relevance to engineering, medicine, management and education. We devote this book series to the communication of recent advances and reviews of revolutionary ideas and their application to practical concerns.

Edward G. Anderson Jr. · Nitin R. Joglekar

The Innovation Butterfly

Managing Emergent Opportunities and Risks During Distributed Innovation

Edward G. Anderson Jr.
McCombs School of Business
The University of Texas at Austin
Austin, TX, USA

Nitin R. Joglekar
School of Management
Boston University
Boston, MA, USA

This volume is part of the NECSI Studies on Complexity collection.

ISSN 1860-0832   ISSN 1860-0840 (electronic)
ISBN 978-1-4899-9988-7   ISBN 978-1-4614-3131-2 (eBook)
DOI 10.1007/978-1-4614-3131-2
Springer New York Heidelberg Dordrecht London

Printed on acid-free paper

Springer is part of Springer Science+Business Media (www.springer.com)

New England Complex Systems Institute

President
Yaneer Bar-Yam
New England Complex Systems Institute
24 Mt. Auburn St.
Cambridge, MA 02138, USA

NECSI

For over 10 years, The New England Complex Systems Institute (NECSI) has been instrumental in the development of complex systems science and its applications. NECSI conducts research, education, knowledge dissemination, and community development around the world for the promotion of the study of complex systems and its application for the betterment of society.

NECSI was founded by faculty of New England area academic institutions in 1996 to further international research and understanding of complex systems. Complex systems is a growing field of science that aims to understand how parts of a system give rise to the systems collective behaviors, and how it interacts with its environment. These questions can be studied in general, and they are also relevant to all traditional fields of science.

Social systems formed (in part) out of people, the brain formed out of neurons, molecules formed out of atoms, and the weather formed from air flows are all examples of complex systems. The field of complex systems intersects all traditional disciplines of physical, biological and social sciences, as well as engineering, management, and medicine. Advanced education in complex systems attracts professionals, as complex systems science provides practical approaches to health care, social networks, ethnic violence, marketing, military conflict, education, systems engineering, international development and terrorism.

The study of complex systems is about understanding indirect effects. Problems we find difficult to solve have causes and effects that are not obviously related. Pushing on a complex system here often has effects over there because the parts are interdependent. This has become more and more apparent in our efforts to solve societal problems or avoid ecological disasters caused by our own actions. The field of complex systems provides a number of sophisticated tools, some of them conceptual helping us think about these systems, some of them analytical for studying these systems in greater depth, and some of them computer based for describing, modeling or simulating them.

NECSI research develops basic concepts and formal approaches as well as their applications to real world problems. Contributions of NECSI researchers include studies of networks, agent-based modeling, multiscale analysis and complexity, chaos and predictability, evolution, ecology, biodiversity, altruism, systems biology, cellular response, health care, systems engineering, negotiation, military conflict, ethnic violence, and international development.

NECSI uses many modes of education to further the investigation of complex systems. Throughout the year, classes, seminars, conferences and other programs assist students and professionals alike in their understanding of complex systems. Courses have been taught all over the world: Australia, Canada, China, Colombia, France, Italy, Japan, Korea, Portugal, Russia and many states of the U.S. NECSI also sponsors postdoctoral fellows, provides research resources, and hosts the International Conference on Complex Systems, discussion groups and web resources.

The New England Complex Systems Institute is comprised of a general staff, a faculty of associated professors, students, postdoctoral fellows, a planning board, affiliates and sponsors. Formed to coordinate research programs that transcend departmental and institutional boundaries, NECSI works closely with faculty of MIT, Harvard and Brandeis Universities. Affiliated external faculty teach and work at many other national and international locations. NECSI promotes the international community of researchers and welcomes broad participation in its activities and programs.

Acknowledgements

Books of this genre are hardly ever the sole work of their authors. We have learnt from, and built upon the thoughts and suggestions of many people. Where possible, we have cited those works that have contributed to our thinking, but there are many others whose ideas, feedback, and support deserve our thanks as well.

In particular, this book resulted from a sabbatical by Nitin Joglekar who visited the University of Texas in the spring of 2005. His visit was sponsored by the Information, Risk, & Operations Management Department of the McCombs School of business, which permitted the authors to collaborate on a day-to-day basis. We would like to thank Anant Balakrishnan, Tom Shively, and the rest of the IROM Department for making this sabbatical possible. Nitin also wishes to thank the Boston University School of Management for this sabbatical opportunity.

John S. Butler, Robert Peterson, and the rest of the University of Texas IC^2 Institute co-hosted our efforts and provided us with office space. They must also be commended for their continued financial support throughout this entire endeavor. Additionally, the National Science Foundation (Award Number SES #0925004) helped support Ed Anderson's contribution to this work as well.

Many people contributed important suggestions to the content of this book. David Angelow, Steven Baggette, Geoffrey Parker, and Scott Palmer provided many conversations that helped us formulate this book initially. John Butler, aside from supporting us with resources, also provided a number of comments about the book that not only improved it but also gave us sage advice on how to navigate the vagaries of the book publishing process. We also owe intellectual debt to several of our teachers, co-authors, and colleagues. On the product development, management of complexity, and innovation physics fronts, we are indebted to Carliss Baldwin, Tim Beck, Tyson Browning, Gary Burchill, Alison Davis-Blake, Janice Carrillo, Steven Eppinger, Lee Fleming, Charles Fine, Cheryl Gaimon, Raghu Garud, John Gray, Stan Gryskiewicz, Wally Hopp, Bala Iyer, Stelios Kavadias, Anil Khurana, Vish Krishnan, Dorothy Leonard, Moren Levesque, Christoph Loch, Kamalini Ramdas, Aleda Roth, Melissa Schilling, Glen Schmidt, Enno Siemsen, Ganesan

Shankar, Ram D. Sriram, Stefan Thomke, Karl Ulrich, Christian Terwiesch, Daniel Whitney, Rohit Verma, and Ali Yassine for sharing ideas and for shaping our thinking over the years. On the systems thinking and system dynamics side, we are indebted to Jay Forrester, David Ford, Alan Graham, Paulo Goncalves, Jack Homer, James Hines, Elizabeth Krahmer Keating, David Lane, Mark Paich, Rogelio Oliva, George Richardson, Nelson Repenning, Peter Senge, Shoji Shiba, and John Sterman. Finally, from a military science perspective, we are indebted particularly to Edward Anderson Sr., Eric Clemmons, and Peter Paret. Nitin is indebted to his parents Smruti and the late Ram Joglekar for lessons in the realm of creativity and writing. Ed is similarly indebted to his mother Joan Anderson. Nitin also thanks his Boston University colleagues Mark Allan, Frederic Brunel, Paul Carlile, Janelle Heineke, John Henderson, Jonathan Hibbard, Jay Kim, Nalin Kulatilaka, Erol Pekoz, Justin Ren, Stephen Rosenthal, N. Venkatraman, Pirooz Vakili, Marshall Van Alstyne, Sean Willems, and George Wyner for shared interests and many productive discussions about distributed innovation processes and projects. Similarly, Ed thanks his colleagues at the University of Texas McComb's School of Business: Rayan Bagchi, Jim Dyer, Leon Lasdon, Kyle Lewis, Steve Gilbert, Jeff Martin, Reuben McDaniel, Doug Morrice, and Sridhar Seshadri.

Yaneer Bar Yam and Dan Braha have been very kind in sponsoring the publication of this book through the New England Complex Systems Institute (NECSI). In particular, Dan in his capacity as the Editor of the Understanding Complexity Series at Springer Science deserves much credit for taking a chance on publishing this book, which is intended for a more popular audience than previous books in the complexity series. He also made a number of important suggestions.

There are a number of friends and colleagues who reviewed an early version of this book and made many helpful suggestions that we have incorporated into the final version. They include: Gopi Bala, Tyson Browning, Aravind Chandrasekaran, Robb Dixon, David Ford, Jonathan Hibbard, Madhav Joshi, Martha Mangelsdorf, Ardhendu Pathak, Rogelio Oliva, Karthik Ramchandran, Scott Rockart, Stephen Rosenthal, Kalyan Singhal, Leonardo Santiago, and Bo van der Rhee. We would also like to thank all our doctoral students and the students in our classes both at Boston University and the University of Texas for all their excellent input over the years. Paula Maute, who provided editorial support, also deserves many thanks for trying to tame the nonlinear thinking of the authors (perhaps a good thing for researchers, but not so good for book authors) into a more coherent narrative.

We also thank David Giber, Scott Cohen, and Aruna Joglekar for dialogues that helped us in reframing the book, in particular about the leadership aspects of innovation portfolio management. We would also like to most sincerely thank Mary Ann Anderson, who read through several drafts and provided many excellent ideas for framing this book as well as for providing an invaluable sounding board whenever we had questions about how industry innovation leaders would likely react to the book's ideas.

Finally, we would like to thank our families and most especially our wives, Aruna and Mary Ann, for all their loving support during this entire process. We dedicate this book to them.

Contents

Part III   Agile and Distributed Leadership

Chapter 1
Mastering the Innovation Butterfly

This book is directed at those who lead the spectacularly risky and complex technical, organizational, and economic challenges of creating product and service innovations. These leaders exist at all levels of a firm including: bench scientists, systems engineers, and product line architects; project managers who worry about day-to-day technical choices; product line planners and business strategists charged with the growth of product portfolios; directors of R&D, supply chains, and customer support; organizational design and human resource specialists; chief information officers, chief technology officers, vice presidents of engineering and marketing; and chief executive officers concerned about the survival and growth of their organization. These individuals, who are often scattered globally across complex innovation chains, make choices that decisively impact the success or failure of their firms' innovation processes. The goal of this book is to help these leaders master their innovation challenges.

The nature of these challenges can be captured by a vivid image suggested by Edward Lorenz, an American mathematician, meteorologist, and a pioneer in the field of complexity science. Lorenz famously gave a talk titled "Does the flap of a butterfly's wings in Brazil set off a tornado in Texas?" He was describing how, in the physical sciences, systems with large numbers of interacting parts can collectively react in seemingly unpredictable ways to very small disturbances.[1] Individual parts of such "complex" systems follow well-defined physical laws. Yet, because of numerous interactions between parts, some unplanned outcomes may be obtained. Feedback between these parts may reinforce such outcomes, and thus the system as a whole may sometimes amplify even minor disturbances (such as a butterfly flap)

---

[1] Lorenz's discovery was initially reported in a 1963 article on a computation model, where the outcome was crucially altered when he entered the decimal .506 instead of entering the full .506127. He later used the image of butterflies to describe the potential for such altered outcomes in a 1972 speech to the American Academy of Advancement of Science.

- Lorenz, E.N.: Deterministic nonperiodic flow. J. Atmos. Sci. **20**(2), 130–141 (1963).

E.G. Anderson and N.R. Joglekar, *The Innovation Butterfly*,
Understanding Complex Systems, DOI 10.1007/978-1-4614-3131-2_1,
© NECSI Cambridge/Massachusetts 2012

to create large-scale effects that are difficult—and in some cases impossible—to anticipate. While Lorenz was describing the behavior of a physical system, these "butterfly effects" or "emergent phenomena" also appear in the realm of human organization.[2] In this book, we adapt these ideas to the sphere of innovation by creating the metaphor of an "innovation butterfly" to describe how seemingly minor decisions, actions, ideas, or events may lead to unpredictable, enormous, and irreversible change from the set plans for all the firms, markets, and people that comprise an innovation system.

As one example of an innovation butterfly creating an irreversible tsunami of change, consider the humble, everyday minivan. For many years, minivans only had three doors. In the front row, the driver and passenger each had a door. The second row of passengers, however, had a single sliding door to their right, but they had no door to their left. So they could only get in and out of the minivan on their right-hand side. All available market research at the time indicated that consumers had no desire for this missing fourth door. But Chrysler ignored this research. Their 1996 design included a fourth door, and customer expectations changed permanently. Now, every firm's minivan must have a fourth door or else fail in the marketplace. But the story of this innovation butterfly does not end there. Despite this dramatic shift in standards, customers' desires continue to ratchet upward. Now *automatic* sliding doors are becoming the norm. In the authors' experience, many children no longer even know how to open an old-fashioned manual sliding minivan door! Who knows what will be next?

Innovation butterflies are everywhere. According to surveys of our students who have worked in innovation settings, most of their teams spent at least *one-third to half of their time* chasing the effects of butterflies—activities that they simply did not anticipate at the start of a typical workday. The reader may verify this by asking themselves: how much of their own time do they spend on activities that they planned at the front end of their day or in the beginning of a planning cycle, especially when they are engaged in an innovative activity? And every one of these butterflies can result in a tsunami of change just like the minivan with the fourth door.

Leading firms successfully in the face of these tsunamis of change unleashed by innovation butterflies is the central challenge facing innovation leaders. This book focuses on helping these leaders attempting to master this central challenge by teaching them to *shape their decisions, both large and small, in the complex innovation system to create tsunamis of positive outcomes for their firms while managing any associated risks in a proactive manner.* This approach provides much greater

---

[2]Caveat: We use the terms "innovation butterfly" and "emergence" interchangeably. These terms are defined in the glossary. In our context, based on the norms of complexity science, emergent means that a system behavior evolves out of initial conditions. There are a number of other definitions and connotations of the term emergence either in the management or in the social science parlance. For instance, even in a traditional project, such a constructing a building, a project manager may face uncertainty, and term the intermediate and final outcomes of the construction process as "emergent." We do not address such alternative views of "emergence."

leverage than focusing solely on the mechanical execution of well-specified plans because the complex, almost "living," nature of the innovation system renders mere efficiency improvements (such as only reducing the time to market without adding significant value through a new product) irrelevant, misguided, or even counterproductive.

In the remainder of this introductory chapter, we begin to address how to master the innovation butterfly by identifying the risks and opportunities associated with managing innovation in a complex, almost "lifelike" system. We discuss how addressing this challenge requires a special form of leadership, beginning with the early detection of patterns of change in the innovation system, also known as "reading the tea leaves," to identify potential innovation butterflies and shape them creatively. To aid this endeavor, we describe a planning process (the *Scout–Roadmap–Orchestrate–Maneuver* cycle) and tools: *Scaling* (to get a larger view of the problem), *Recognizing* (looking for butterflies while they are still controllable), *Agile Portfolio Planning* (proactively adjusting the product roadmap to shape the innovation butterfly), and *Distributed Leadership*. We then show how these four key ideas operate in a real-life innovation setting: the video game industry. Finally, we outline the remaining three sections of this book, which expand on the key ideas introduced in this chapter by using evidence from various settings—ranging from the automotive to medical device industries—to describe how these challenges evolve and how to respond to them.

As a pointer before we begin, this book draws upon a number of terms such as "emergence" and "scaling." Instead of formally defining these terms, we have tried to provide simple explanations when these terms first arise. However, the reader may also refer to the glossary in the appendix where they are more formally defined.

The Central Challenge: Managing "Living" Innovation

The ideas that we describe in this book can apply to structured tasks such as the construction of a house. However, we have focused on product and service innovation projects (or portfolios) because they are, by definition, about doing something novel. Customers of new products hope that innovations, large and small, will solve their problems and perhaps even surprise and delight them by solving other problems they did not even know they had. Firms engaged in innovation hope to anticipate these demands, which are forever ratcheting ahead based on innovations already released to the market. Innovation workers hope to learn how to better create these innovations, often through trial and error. Delivering on these three hopes involves many uncertainties, risks, and iterations. The result is an innovation system that often produces an endless stream of surprising events and outcomes and acts in many ways as if it were "alive." The lifelike nature of the innovation system makes it a breeding ground for innovation butterflies. Some of the negative outcomes resulting from early decisions creating butterflies long consequences have been likened to the actions of HAL 9000 computer in Stanley Kubrick's film 2001: A Space Odyssey.[3] For example, a small change in a software configuration once created a bug that brought the whole European space program down to its knees.[4]

As Christopher Langton and a host of other physicists have established, butterfly effects in complex systems can mimic lifelike behavior.[5] Seemingly small changes created by executive decisions, market trends, design choices, or a myriad of other factors result in innovation butterflies whose fluttering effects can cascade into tsunamis of change that can either benefit or destroy a system. Planning for and shaping the course of these tsunamis, or their more technical name, "emergent phenomena," is the central management challenge in innovation systems.

The impact of most innovation butterflies is relatively innocuous. They hinder innovation teams by creating unplanned tasks that might take 5–10% extra resources. Innovation butterflies can also hamper the focus and productivity of innovation workers, causing delivery delays and cost overruns in the range of 10–30%. However, this is only the tip of the iceberg. The innovation system can sometimes reach a "tipping point" in which project management becomes dominated by handling escalating customer demands for unplanned features.[6] And it can get even worse. Over the course of the development and marketing of multiple incremental innovations, the effects of innovation butterflies can accumulate, consume enormous resources, and create major disruptions that, when not shaped carefully, pose a much greater threat. Industry is littered with the road-kill of companies that could not cope with innovation butterfly effects. Polaroid Corporation is a classic example of a firm that successfully capitalized on an innovation (the instant photo camera) but could not manage to make the next technical transition (to digital photography). Blockbuster Inc., the one-time titan of the video rental business, successfully navigated the technological transition from videocassettes to DVDs. However, the market shift to DVD rental via mail, pioneered by Netflix, followed by the market shift to on-demand digital movie downloads has lead Blockbuster to filing for bankruptcy in 2010. It has since been acquired by satellite TV provider Dish Network for only a fraction of its peak market value.

---

[3] See http://www.imdb.com/title/tt0062622/; also see Arthur C. Clark's (1968) novel.

[4] Dowson, M.: The Ariane 5 software failure. Softw. Eng. Notes **22**(2), 84 (1997).

[5] Langton, C.G.: Life at the edge of chaos. Artificial Life II **10**, 41–91 (1992).

[6] Such situations result in severe questioning of initial plans, and they sometimes lead to litigation around assignment of responsibilities for the escalation of scope, delays and allied cost overrun. See:

- Cooper, K.G.: Naval ship production: a claim settled and a framework built. Interfaces **10**(6), 20–36 (1980). Special Practice Issue.
- Peterson, J.H.: Big dig disaster: was design-build the answer? Suffolk U. L. Rev. 909 (2006–2007).

Why did neither Polaroid nor Blockbuster get on the digital bandwagon when the early signs of such disruptive innovations were apparent? The reason is that the trickiest part of the challenge of managing "living" innovation systems, and the butterflies that they produce, lies in constantly looking for patterns at the system level and then rapidly shaping decisions to exploit the patterns or the "tea leaves" before the system changes beyond recognition. Recognizing that this is necessary, however, and actually making rapid informed decisions, are two different things. Pattern recognition based on early and emerging information and gearing an entire organization for rapid action based on such information is a difficult management challenge, particularly for firms like Polaroid and Blockbuster that had been so successful.

Looking for Patterns and Shaping the Innovation Butterfly's Path

If an innovation leader can understand the drivers of innovation butterfly effects *and recognize the early patterns of emergent change in a system,* the lifelike nature of the interactions within an innovation system become gradually transparent and, ultimately, shapeable. For example, Apple Corporation began with tremendous success with products such as Apple II early in its history. Then, it faced a series of difficult choices, as some of the follow-on products such as its initial handheld product, the Newton, failed. However, Apple has continuously tried to learn from these signals and adapted to change. Thus, through innovations such as iTunes, Apple Corporation was able, despite a number of tremendous hurdles, to get back into the handheld market and go from success to success with the iPod, the iTouch, the iPhone, and now the iPad. In fact, Apple has been so successful over a sustained period of time that it is actually shaping the trajectory of its innovation butterfly with respect to the digital delivery mode of media in a way that favors Apple.

How does one identify patterns and shape the path of the innovation butterfly? Consumer research is helpful but often does not get to the heart of these matters. Determining current consumer needs is difficult enough, but trying to estimate where these needs are going (reading the "tea leaves") and then shaping them is even trickier.[7] However, reading these tea-leaves opens up numerous options. *This means that, not only must innovation leaders constantly "read the tea leaves" for what is likely to transpire in the future, but also must do so for a number of potential scenarios. They must then somehow maneuver their product portfolios, and the entire industry ecosystem around it, toward favorable outcomes for their firms.* The ability to respond appropriately to these options is determined by whether the

---

[7]This has been pointed out by Garud, R., Karnoe, P.: 'Path Creation as a Process of Mindful Deviation'. In Path Dependence and Creation, R. Garud and P. Karnøe (eds.) Lawrence Erlbaum Associates: pp 1–38.

innovation organization can make continuous adaptations in their decision making as the innovation system evolves. Facilitating these types of adaptations requires a new proactive style of leadership.

Leadership Opportunities in the Innovation Butterfly Age

Aside from describing the genesis and growth of innovation butterflies, a second goal of this book is to provide, at a nuts and bolts level, ideas and tools that innovation leaders can use to better read the tea leaves associated with innovation butterfly effects and then respond to them creatively and powerfully. Because these effects result from complex interactions among numerous small and large decisions, we draw upon established tools and results from complexity science and apply them to innovation systems. Our ideas are also heavily influenced by military science, whose goal is to create organizations that thrive on literal chaos and complexity, in order to exploit them to the organization's advantage. We believe that the central goal of leaders in innovation firms is to seek out and exploit the outcomes of innovation butterflies. To do so effectively requires rapid collection of data, ongoing assessment of potential risks and benefits, and continual adaptation and reworking of the initial plans. We believe that these activities will deliver far superior product and portfolio performance than routine improvements during individual project management.

The keys to effective leadership in the age of the innovation butterfly are fourfold.

- First, leaders in an innovation firm must "*scale*" their view of the problems to facilitate pattern recognition. That is, they must shift from focusing on managing individual tasks or projects to managing the entire portfolio of the firm's innovation projects. Such a scaling process allows a leader to focus on and effectively track a select few parameters.
- Second, innovation leaders must learn how to analyze the marketplace response to their own (or their employees') decisions as well as promote the development of their employees' capabilities. By doing so, innovation leaders can *recognize innovation butterflies while they are still within their area of control*, before their effects become too large to manage proactively.
- Third, innovation leaders must conceive of *innovation planning as a real-time process consisting of agile, adaptive cycles of portfolio planning, data collection, and maneuvering*, rather than as an annual planning exercise. In other words, when the current plan is underway, planning for the next maneuver to exploit the innovation butterfly should also be in progress.
- Finally, innovation organizations must *foster an empowering, decentralized culture and leadership practices* so that individual innovation leaders can execute and adapt these maneuvers to gain the maximum effect.

Before discussing the appropriate concepts necessary to support these key ideas in detail, let us consider a case that illustrates some of the drivers underlying the innovation system in which they must operate.

An Agile Chase: The Innovation Butterfly in the Videogame Industry

To describe the innovation butterfly in greater detail, we use the videogame industry to illustrate certain ideas developed in the field of complexity science. We then describe how innovation systems are similar and different from complex physical systems. Based on this knowledge, we then ask how an innovation leader might manage, and even leverage, the innovation butterfly.

The U.S. video game industry has outpaced Hollywood Box Office sales since 2004 and overtook the music industry's CD, downloads and vinyl records sales in 2008. However, unlike Hollywood movies which grew over a 50-year period, with little competition from alternative technologies, the video game industry's rapid growth has occurred at precisely the same time as great uncertainty regarding emerging technologies such as 3D images and allied consumer preferences are threatening this industry.

Up until 2007, the leading console makers in the video game industry—Nintendo, Microsoft, and Sony—came out with a series of game consoles that increased computational power to provide ever more realistic video images and sound effects. Each new console released resulted in escalated customer expectations of greater realism. A core group of the customers were young males, traditionally identified as "hardcore gamers" who grew up playing on such consoles, and had increasing sophisticated know-how and expectations about the graphics technologies that rendered life like images. However, in 2007 Nintendo Inc. replaced its venerable GameCube with the Wii. The Wii attempted to reshape market expectations by radically departing in design from its closest competitors' games—Microsoft's Xbox 360 and Sony's PlayStation 3. The Xbox 360's and PlayStation 3's designers had maintained the traditional trajectory of game console design by implementing greater sophistication in graphics and sound with each new generation of computational technologies, for example, going from 8-bit to 16-bit processors allowed the designers to offer a much faster rendering of finely textured images. Wii designers did something different by attempting to expand the demographics of the video game market beyond the current "hardcore" player demographic.[8] In particular, they wanted to expand their market to the entire family, including girls, women, the 55+ age group, and "casual" gamers, who might not want to dedicate the hundreds of hours necessary to master the traditional multiple joystick-and-buttons game controller for just a couple of hours of fun. To this end, they created a new type of game controller, the "Wiimote," based on accelerometer technology that measures the movement of the player's hand, rather than relying on the traditional joystick. The aim of Nintendo's designers was to create a more intuitive feel and control of the game,

---

[8] Much of this evidence is documented in a set of interviews by S. Iwata, President of Nintendo. Two interviews were particularly helpful: Iwata asks: The Wii Remote. http://us.wii.com/iwata_asks/index.jsp. Accessed 13 June 2011 and Iwata asks: The Wii Hardware. http://us.wii.com/iwata_asks/index.jsp. Accessed 13 June 2011.

which they did quite successfully. Thus, when a Wii player wants to hit a ball with a "racquet" in a tennis video game, he simply swings the hand that holds the controller as if he were swinging an actual racquet instead of manipulating a joystick or pushing a button.

At the same time, the Wii's designers also made the critical decision *not* to significantly improve the graphics and sound performance over the previous generation of game consoles because they felt that the market had reached a point of diminishing returns in what customers, especially new video game customers, were willing to pay for improved graphics and audio realism. Their decision had been informed in part by the fact that sales of Sony's PlayStation 2, the less graphically realistic predecessor of PlayStation 3, continued to exceed that of its "improved" successor through mid-2008. In essence, Nintendo was trying to shift the direction of consumer expectations onto a new course based on realistic motions rather than based on visual realism. We now review this case in light of the four keys to effective leadership in the age of the innovation butterfly.

Scaling

Nintendo's Wii choice was based on its recognition through market analysis that the old industry measure of superior product portfolio performance—the realism and clarity of video quality and sound—was reaching a point of diminishing returns. In particular, the market of traditional "hard-core" gamers, which was overwhelmingly male teenagers and young adults who played 20–40 hours per week, was saturated. Therefore, Nintendo "scaled up" their own view of the market place and explored what was the value being offered to these "hard-core" gamers, and asked if there were any alternative ways of getting around this market saturation. For example, the relevant technical performance measure assigned to the engineering teams could be changed from graphics fidelity to motion-sensing fidelity. Then, it asked its software application developers to write gaming application using these motion-sensing capabilities. Finally, Nintendo retargeted its marketing plans toward a new market segment. That is, through a series of small choices, based on development of sensor technologies, Nintendo created a new and differentiated measure of product portfolio performance—the realism of the motion-sensing by an input device—which could engage previously ignored market segments such as women and "casual gamers" with a novel group of games based on more realistic motion-sensing controllers.

Recognizing Innovation Butterflies While They Are Still Controllable

Nintendo's decisions proved to be wildly successful. The Wii today has the largest base of videogame console purchasers among its competitors by a wide margin. It achieved this by appealing to a wider demographic base, just as it had planned.

The fastest growing demographic group of video gamers in the USA are players over 60 years old, who are attracted in part by Nintendo's placement of Wii consoles in senior citizen centers and retirement homes around the country. In order to appeal to this "senior" segment of consumers, Nintendo has created novel uses of the Wii, such as simulated bowling tournaments.[9]

While a tsunami of change was hitting the video game market place, could Nintendo control and shape what came next? Many third-party game developers, who all video game console companies count on to generate royalties and boost future sales, initially found it difficult to develop games for the Wii because of the novelty of the motion-controller. These developers had been working on sophisticated 3D graphics and had little or no experience with programming to respond to motion-sensor-based controllers like the Wiimote. Moreover, many industry insiders and traditional "hard-core" gamers are quite critical of the Wii, asserting that either it is no different from the graphics-based game consoles of a generation earlier or that the Wii's remote controller is nothing but a passing fad just like the joy sticks of earlier generation of games.

Nintendo's Wii planners provided for such contingencies. In particular, their decision to keep improvements in the Wii's graphics and sound to a modest level held down development costs for third-party software developers. This has enabled these developers to create games for the new controller at a fraction of the cost necessary to create games for the Wii's competitors, which helped boost the number of games that can be played on the Wii far above its competitors. This has proven critical, because many customers buy game consoles depending on what and how many games are available for that controller. Another successful strategy of Nintendo's innovators was to direct their own game development capabilities toward providing a bundled set of games called Wii-sports that offers easy-to-play (thanks to the Wiimote) electronic versions of established games such as bowling, golf and tennis. This primed the pump for Wii demand, because the intended customers were already familiar with these sports.

Agile Portfolio Planning and Execution (or, Maneuvers by Nintendo and Competitors)

Once Nintendo innovation leaders realized that they had begun to attract new demographic groups to their market, they rapidly moved to exploit their success by continuing to rely on their internal capabilities to create innovative video games such as Wii-Fit. The Wii-Fit uses yet another intuitive controller, a balance board, as an input to games to encourage physical exercise, such as yoga, aerobics, and even simulated ski-jumping. This has proven exceptionally attractive to women, a generally neglected gaming demographic prior to the Wii-era.

---

[9] Conway, L.: Why Senior Citizens Love the Wii. http://www.npr.org/blogs/thetwo-way/2009/07/why_senior_citizens_love_the_w.html (2009).

Furthermore, to increase their lead in realistic input devices, Wii developers added a second motion controller, the Nunchuk, enabling both hands to make separate inputs. They then followed this up with several other extensions such as the Wii Zapper, which looks like and creates the virtual functionality of a ray gun, and the Wii Wheel, which does the same for video game automobile steering wheels. Finally, they followed up most recently (2010) with the Wii MotionPlus, which utilizes a tuning-fork gyroscope to enable even better and faster rendering of the Wiimote controller motion on the video screen that makes the motion experience even more lifelike for gamers

Culture: Empowering Creative Leadership

Our familiarity with the culture at Nintendo, and video gaming sector in general, is admittedly limited because much of the information comes from publicly available sources rather than interviews and direct observations. However, it is clear that Nintendo values creativity and empowers creativity through distributed leadership. One of the Wii's creators commented when discussing the creation of the Wiimote controller, "Nintendo is a company where you are praised for doing something different from everyone else. In this company, when an individual wants to do something different, everyone else lends their support to help them overcome any hurdles. I think this is how we made the challenge of the Wii a possibility."[10] Research indicates that successful management of innovation workers typically follows this attitude.[11] Moreover, our own field research in the videogame, automotive, and energy industries indicates that a decentralized approach is much more successful in innovation settings because local decision-makers "closer to the ground" can make much more rapid and effective adjustments to unpredictably changing conditions than can a top-heavy, centralized, bureaucracy.

Lessons from the Wii: The Innovation Butterfly Never Sleeps

Recent trends indicate that the Wii may have created deep shifts in the trajectory of the gaming industry. Both Sony and Microsoft have recently come out with more intuitive controllers that use technology similar to the Wiimote. These, in turn, will likely create even more customers who are receptive to the use of more intuitive, accelerometer-based controllers. At an even deeper level, the Wii may have unleashed

[10]Quote from Genyo Takeda in Iwata, S.: Iwata Asks: The Wii remote. Retrieved from http://us.wii.com/iwata_asks/index.jsp. Accessed 13 June (2011).

[11]Glen, P., Maister, D.H., Bennis, W.G.: Leading Geeks: How to Manage and Lead People Who Deliver Technology. Jossey-Bass (2002).

a further innovation butterfly by creating a climate in which designers are becoming ever more capable in designing games for accelerometer-based game controllers which the Wii promotes. At the same time, consumers are becoming more demanding while the novelty of motion controller wears thin. Thus, the future of the game industry is clouded with uncertainty. However, two important points of the Wii's development stand out. One is that an innovation can deliberately shape the trajectory of market expectations. Furthermore, building capabilities to adapt to these changing expectations takes time, but once that adaptation occurs, it can create follow-on shifts in the evolution of market expectations (e.g., the rise of games designed for smartphones). The second point is that Nintendo followed a strategy that leveraged its capabilities in both a proactive and adaptive manner. Its innovation leaders clearly understood the complex nature of the change they were trying to effect and its associated risks. They protected against these risks by first leveraging their capabilities in traditional games by, for example, building on their Mario series software franchise by releasing Mario Smash Brothers games and secondly by reducing costs for third-party developers by keeping the improvement in console graphics minimal. This protected Nintendo in case the new controller did not catch on quickly or proved to be a passing fad. However, once the new controller caught on and opened new demographics, Nintendo adapted to capitalize on these changes by developing new products tailored to these new demographics such as the Wii-Fit. In a similarly adaptive fashion, Apple then jumped on the bandwagon by incorporating its own intuitive user input capabilities into its iPhones and iPads and then created the infrastructure necessary—often through adapting their existing capabilities—to exploit the new demographics of the game market and expand them further.

Thus, the Wii case underscores the keys to leadership in the age of the innovation butterfly: Nintendo's innovators scaled up their view of the market to enable pattern recognition at the portfolio level (i.e., the consumer shift away from complex graphics to simpler motion-based controllers); they recognized the evolution of innovation butterflies (i.e., low cost development paradigm for development by deemphasizing expensive graphics attracted software suppliers, which flooded the market with novel and interesting games); they performed real-time, agile product portfolio planning and maneuvering; and they empowered and enabled creative and rapid decision making by their innovation employees. We now briefly outline the know-how, tools, and corporate culture that can provide and nurture these leadership keys.

Know-How, Tools, and Leadership

Thus far, we have described basic ideas on know-how and leadership via the Wii case and other simple examples as an overview of a strategy for coping with the innovation butterfly. In the remainder of the book, we use several detailed examples from multiple industries to provide a fuller picture of each of these concepts and we describe their application and managerial implications in greater depth.

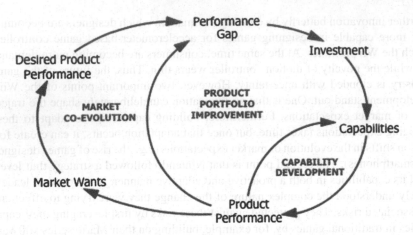

Fig. 1.1 An innovation system with multiple feedback loops. The escalation of expectations in the innovation system derives from three nonlinear feedback loops, which render the system dynamically complex. This causes the innovation system to behave predictably some of the time and almost randomly at other times. Moreover, shifts between predictability and unpredictability can occur unexpectedly. This results in a need for both a proactive and reactive agile planning process. This figure has three loops. How to read the loops and their outcomes are further discussed in Chap. 2

These examples, as well as the rest of this book, are divided into three sections: understanding complexity, agile tools for managing complexity, and distributed leadership.

Understanding Complexity and the Evolution of Innovation Butterflies

When we discuss the Wii case, we deliberately concentrate upon the innovation butterfly unleashed by the Wiimote controller. However, other potential innovation butterflies existed at the time. Researchers in innovation such as Clayton Christensen[12] have pointed out that disruptions exist in innovation systems. We view these disruptions as tsunami-like outcomes and address how to manage the effects of the innovation butterfly both before and after the tsunami has occurred. After all, *recognizing that tsunamis can occur* is not enough to bring about innovation—*understanding how they occur* is imperative. As shown in the Chrysler minivan example, the very act of introducing new products shapes the market's tastes and landscape, which then influence firm's future product decisions. This leads to a continuous *Escalation of Expectations.*

---

[12]Christensen, C.: The Innovator's Dilemma. Collins Business (2003).

As stated earlier, innovation systems are complex because innovation influences expectations, which in turn influence the planning of new innovations. Further, by carefully planning and developing new innovations, firms create new capabilities, which can stimulate the creation of additional innovations. These effects are illustrated in Fig. 1.1. The most important point in this figure is that the three loops, "Market Co-Evolution," "Product Portfolio Improvement," and "Capability Development," feedback on each other. For example, in the market co-evolution loop, if *product performance* improves along some dimension, such as when Wii's controller became more intuitive, it creates *market demand* for intuitive controllers and more games that can use those appealing controllers. Thus, changed *market demand* influences future *desired product performance.* This creates a *performance gap* that drives *investment,* which, over time, will improve a firm's *capabilities* and ultimately its *product performance,* completing the loop. Similar feedback loops occur through a firm's product portfolio and capability development loops.

Each of these loops has delays. That is, it takes time for a change in any variable such as *market wants* to fully influence other variables. Moreover, the changes are nonlinear; that is, an increase in *investment* may not necessarily directly and proportionally affect a firm's *capabilities.* A number of important implications flow from the fact that the innovation system has multiple, nonlinear feedback loops, but only two implications concern us here. One is that these systems are dynamically complex. In practice, what this means is that the innovation system may behave in a reasonably predictable manner most of the time. However, it can also be pushed into behaving chaotically, in which case any deviation from plan resulting from external influence or internal change, no matter how small (e.g., the smallest current created from a butterfly's flutter) can set the system off onto an essentially unpredictable trajectory. This is what gives the escalation of market expectations its lifelike nature, and this is why a firm must both *proactively and reactively* plan, because the system can switch between predictable and unpredictable states at any time.

Going back to the Wii case, the Wiimote controller shattered marketing conventions as to what a video game controller and even a videogame should be. And, it created a fresh infusion of customers (e.g., the 55+ age group), who now demand unique features, such as slow-speed work out capabilities, from these games. Furthermore, Nintendo developed the capability to satisfy wants for videogame customers that it did not even know existed. This is most clearly seen in the Wii-Fit market segment for Nintendo, which is connecting the healthcare industry with the gaming industry to some extent. Will the Wii continue this trend by bringing other nontraditional fields into the video game realm? If so, which ones? How can Nintendo, or any other innovation firm, recognize what shifts in market expectations might play out, and in what time frames? More fundamentally, what creates the complexity of an innovation system and how do the effects of an innovation butterfly play out? We discuss these questions in Chap. 2, while describing an "Escalation of Expectations" Principle, and offer case evidence based on the 1973 U.S. Clean Air Act and how it helped create computer-controlled automobiles, including such advances as antilock brakes, smart airbags, and integrated navigation systems.

Another important property of complex systems is that they often result in useful, yet unintended and difficult-to-foresee consequences. This is summed up in the *Principle of Exchange*. The principle was crystallized for us by the energy expert Geoffrey Parker during a conversation about Hurricane Katrina: "In complex systems you never solve a problem fully; you just end up swapping one set of problems with another. Hopefully, you end up with a better set of problems." For the Wii case, the most potent butterfly that evolved from the Wiimote controller's development is the subsequent casual gaming culture involving new types of customers such as the 55+ age demographic into an industry that had been accustomed to serving hardcore gamers representing a much younger customer demographic. Furthermore, Nintendo's competitors and third-party game developers have been learning how to develop games using these new types of motion inputs. Hence, the Wii may have inadvertently opened the gaming industry to an invasion by accelerometer-equipped smartphones like the iPhone that run game apps. How could Nintendo have recognized this threat ahead of time? We explore the ramifications of the Principle of Exchange in Chap. 3 using the examples of river engineering in which any given fix seems to create another problem, and Exubera, an inhaled form of insulin that initially held much market promise but managed to implode anyway.

Finally, it has been known since at least the time of Adam Smith, the 18th century economist, sometimes identified as the Father of Capitalism, that the most effective method to manage complex systems is to decentralize their management to do away with issues such as ineffective bureaucracy on top. While this is indeed the case (e.g., consider the fate of the centrally planned economy of the Soviet Union), such decentralization also creates two major problems. One is how to coordinate decentralization in an effective manner so that innovation leaders are not pulling in different directions based on their individual perceptions of what is important for their parent firm. The other is to keep the decentralized innovation employees and subordinate managers from acting in their own self-interests. This leads to the principle of *Providential Behavior*. That is, if employees pursue their self-interests, and they often do, dysfunctional outcomes such as empire building, personal rivalries, look-alike products within the firm's portfolio, and pet projects emerge that waste the parent firm's precious innovation resources.

Returning to Nintendo's Wii case, much of this dysfunctional behavior seems to be absent. However, what are the obstacles that Nintendo faced—and must continue to face—in creating an empowered, decentralized, yet team-centered culture? These questions are discussed fully in the Principle of Providential Behavior in Chap. 4 and a case concerning software specification and estimation at SofTex, a global software company.

Agile Tools for Managing Complexity

We now briefly describe the tools that enable agile management of project and portfolio complexity: information scaling through analytics, planning using maneuver-oriented competition, and modularizing risk.

Information Scaling Through Analytics

The key change at Nintendo that was responsible for the Wii was the realization that ever-greater visual fidelity was reaching the point of diminishing returns with respect to growing the market. Therefore, they took a higher level or scaled-up view of the consumer's need and decided that the space to focus on was motion capability rather than the visual fidelity. Like Nintendo's leaders, the first problem an innovation leader at any firm must cope with is how to read the "tea leaves"— that is, how to scale up their view of the problems. They then must decide on which types of data to gather (and how frequently) in order to monitor multiple scales of information at the portfolio and project level. Over time, with the advent of information technologies, the amount of available data, especially in distributed (e.g., across multiple countries) innovation settings, is growing almost exponentially. The development of innovation analytics and the related business analytics capabilities that can turn these data into useful information to guide innovation leaders and their staff are still in its infancy.[13] However, a few facts are known. One is that consistently successful firms in innovation management such as Apple (currently) and Sony (in the 1980s and 1990s) are often notorious not only for their individual project successes, but also for the cumulative nature of those successes. What is often absent from the analysis is the many individual project failures (e.g., Apple's Newton) that each firm experienced during these periods. However, even those failures provided valuable information for the development of subsequent projects. What is required for successful innovation analytics is a shift of focus analogous to anticipating the movement of a *herd* of cattle, rather than focusing on the movements of individual cows. Cowboys make this mental shift, from watching the movements of individual cows to watching the aggregate movement of the herds, as an everyday part of their jobs. Moreover, experienced hands learn general principles about herd mentality—e.g., cattle typically move more easily across flat areas and move most easily downhill (which can be either helpful or hindering, depending on the ultimate destination).[14] To make the metaphor accurate for innovation leaders, however, imagine that one can only see the herd through an intense fog and must predict the herd's movements several hours into the future. This is the challenge of information scaling as faced by the innovation leader. Once this skill has been mastered, however, an opportunity opens up because complexity science researchers have shown that the behavior of complex systems can often be described by a few simple rules or principles (like cattle herds tend to move downhill). Thus, by scaling their attention appropriately, innovation leaders can begin to make some directional predictions about the behavior of the innovation system.

---

[13] Davenport, T.H., Harris, J.G. Competing on Analytics: The New Science of Winning. Harvard Business Press (2007).

[14] http://www.rustyspurr.com/TeamBuildingFocal.html.

Yet, how exactly do innovation leaders scale information to the portfolio level from the detailed data available at the project level? The answer remains somewhat unclear, particularly as it will vary significantly across industries. Nonetheless, some elements are clear enough. For one, the innovation system, which is complex, may, over the short run become predictable. Thus, paying attention to the trajectory, not just the current state of the market landscape, and tracking competitor product portfolio pipelines, will yield important information dividends in creating useful information analytics for the innovation leader. Another necessity is to closely monitor the state of competitor's products (trend in the product mix, market share, perceived profitability, etc.) because they influence customer expectations and hence shape the evolution of a firm's own project portfolio. In fact, the evolution of competitor's portfolio, and allied customer responses, often prove the best indicators of where innovation systems arc headed: for instance, is there a predictable pattern? Or is the system disrupted, or even becoming chaotic?

As innovation leaders develop innovation analytics capabilities at the portfolio level, they must also keep an eye on the firm's capabilities and the state of its projects and personnel (i.e., number of projects, target markets, and the types of expertise in house, and in the supply chain). That is, leaders must pay attention to individual cattle as well as the herd; for example, one particular cow's movement toward a cliff could stimulate a stampede. Or, a key designer leaving the team could send a signal for impending changes to an entire team.

One highly successful method to accomplish this task derives from the military's concept of the directed telescope,[15] which places staff officers in a battle's forefront with explicit instructions to rapidly feed information on certain critical issues from the front lines directly to the overall commander, bypassing the normal military chain of command. Interestingly, such a structure is used in many software firms, in which a manager or a key software designer acts as a stakeholder for and conduit of information to upper management for individual projects. It is extremely important to note, however, that these management representatives do not directly lead projects. But they are assigned to roles that allow the access to rapidly evolving data. They do not flood the higher level managers with all these data, because these representatives understand scaling issues. Instead, they only convey information that they deem to be important enough to deserve a higher level attention.

Going beyond the Wii case, how should a firm in other industries determine what parameters to track? How should it monitor their capabilities? How often and by what means should they update their information? Gathering and analyzing such information must be an ongoing task assigned to a team that works for the innovation leader. We examine these issues in more depth in Chap. 5, Maneuver-Driven Competition, using the case of a successful video game design firm Online Alchemy.

---

[15] Van Creveld, M.L.: Command in War. Harvard University Press (2003).

Planning Using Maneuver-Oriented Competition

Recall that when Nintendo recognized that improved video fidelity did not bolster sales, they redirected their innovation skills to improving the intuitiveness of the video game input device. Furthermore, once this began to succeed, Nintendo began to leverage their new capability with input devices to extend the Wiimote with the Wii Zapper, a gun like control stick that allows players to zap their way to score points, and the Wii Wheel, to mimic an automotive steering wheel. Introduction of these devices brought out new types of applications, such as more intuitive automotive racing games, invented by the supply chain partners. Thus, successful innovation leaders must not only gather information on the market landscape, their own capabilities, and competitors' product portfolios, but they must also adapt their planning and management structure to that information and prepare for unpredictable contingencies such as a partner creating a new game niche This is difficult because most organizations manage by predicting only one likely outcome, to which they then focus all the firm's capabilities to optimally exploit. However, if reality deviates from this single predicted outcome, planning must begin again from scratch. This might require a long period of adjustment, particularly if new capabilities, which primarily require trained personnel, must be created. Fortunately, while uncommon in the product innovation field, methods for creating such capabilities can be drawn from fields such as physics, military science, software project management, and organizational science, to enable innovation organizations to thrive in complex systems. In Chaps. 5–7 of the book, we draw upon theories of adaptive leadership from these fields to describe three innovation planning tools for thriving in the midst of complexity: *maneuver-driven competition, modularizing risk,* and *plug-and-play processes.*

More specifically, how can innovation leaders, their staffs, and employees plan proactively while remaining able to exploit potential innovation butterflies? We draw upon the agile project planning methodology from the software industry for inspiration.[16] However, while agile planning has a good track record at the project level, it has been applied much less frequently at the portfolio level. We suggest how the underlying idea of agility can nevertheless be adapted both for planning at the project portfolio level and for applying it to settings beyond the software industry. We do so by borrowing principles from the military and complexity science fields by introducing the concept of *maneuver-driven competition* to adapt agile planning to the portfolio level. Maneuver-driven competition describes an endless Scout–Roadmap–Orchestrate–Maneuver (SROM) cycle in which innovation leaders (1) *scout* the market, as well as the firm's and its competitors' current capabilities using innovation analytics, (2) plot a technology *roadmap* including the capabilities necessary to achieve it, (3) *orchestrate* the firm's actions at subordinate levels by adjusting their

---

[16]The particular challenges of videogame development and how agile development processes are suited to meeting them are discussed further in Chap. 5.

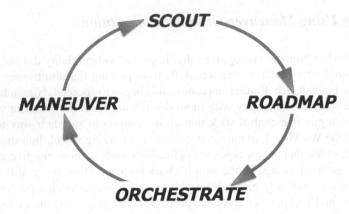

Fig. 1.2   Scout–Road Map–Orchestrate–Maneuver (SROM) cycle

objectives to account for the changes in the road map, and (4) *maneuver* by allowing subordinate innovation leaders to guide their teams forward on their projects for a short time prior to beginning the four-part cycle again. Hence, the innovation leader continually revises her or his plans even before all of the actions from the previous set of plans have played out. This planning results in a continuous cycle of short maneuvers across shifting currents, which requires a balance of market forecasting, planning, and execution and creates a competitive strategy driven by a sequence of rapid maneuvers, or business actions, to adjust to changing conditions.[17] To the extent that a firm can speed up the tempo of this SROM cycle, it can better exploit opportunities resulting from the innovation butterfly than its competitor

The SROM cycle in Fig. 1.2 may look familiar to some readers. It has descended from the Shewart Plan–Do–Check–Act (PDCA) cycle popularized by W. Edwards Deming in quality management and the Observe–Orient–Decide–Act (OODA) loop developed by the legendary Colonel John Boyd of the U.S. Air Force for military command-and-control decisions. (These sources will be discussed more in depth in Chap. 5.)

Adopting a maneuver-driven competition plan, however, is not without challenges because it demands a major change in traditional business practices. For example, scouting is essentially the gathering of information analytics as described earlier; however, the key to its successful application in the SROM cycle is that scouting must become an ongoing process, rather than a perfunctory annual exercise. Further, it is necessary to triangulate observations from multiple

---

[17]Our view builds on Thomke's (2003) work that makes a case for systematic experimentation and learning during innovation processes. We add the elements of adaptive search and distributed leadership to this view of innovation.

- Thomke, S.: Experimentation Matters: Unlocking the Potential of New Technologies for Innovation. Harvard Business School Press (2003).

sources whenever possible, because observations of market trends, product pipe-lines, and especially firm portfolios can minimize these observations' inherent uncertainty. The more accurate and timely a picture that scouting produces, the more rapidly an innovation firm can execute the SROM cycle. The more rapid the SROM cycle, the more agile the firm will become. Otherwise, executing the SROM cycle rapidly without timely and accurate scouting becomes like the cowboy mentioned earlier who is trying to herd cattle through a dense fog. If he drives the herd too quickly through a dense fog, he could cause the herd to stam-pede into an unseen river or over a cliff. In the same way, reacting too quickly to uncertain information can lead to a critical mistake in the use or development of firm capabilities.

Roadmapping also has its challenges. Traditional technology planning first deter-mines the most probable scenario likely to occur, and then plans only for that sce-nario. However, this is not an effective strategy in a complex system because even when the innovation system is behaving relatively predictably, multiple scenarios may still be plausible and they must all be explored and planned for. Plus, the sys-tem may become completely unpredictable at anytime, so roadmaps that have built-in buffers to adapt to unanticipated scenarios must be developed.

Orchestration in a maneuver-oriented competition firm, in contrast, is somewhat simpler than in the traditional innovation firm because the innovation leader and her staff do not need to monitor and intervene in individual projects on a day-to-day basis. In fact, such meddling is counterproductive. A better analogy for the innovation leader to follow is that of a symphony conductor, who does not play any of the instruments in the orchestra. Nor does the conductor instruct the orchestra on how to play the instruments. Instead, the conductor guides the musicians in beat, tempo, and sound to bring the best out of the orchestra. In an innovation firm, this requires individual project managers to specify the goals, timing, and cost of a project and to gently shape the overall outcome with support when needed. Individual project managers can and will take it from there. Innovation leaders must additionally provide project managers with the intent of the project. For example, if it is more critical that a project be executed to develop a needed capa-bility than it is to generate huge revenues from the project, the project leader can then make the appropriate tradeoffs at a local level without losing time by consult-ing with upper management.

This raises the final challenge for the innovation leader, which usually occurs during the maneuver phase. Innovation leaders must avoid the temptation to inter-fere in individual projects; otherwise the benefits of decentralization and a rapid SROM cycle are lost as manager's waste time (waiting for upper management's input) and energy (trying to incorporate that input, which may not even be useful, into the project). Unfortunately, avoiding such interference is particularly chal-lenging as it requires that management act against their intuition to meddle, and it goes against the preached management practice in many traditional firms. We examine how to cope with these issues in Chap. 5: Maneuver-driven competition using case studies of Online Alchemy and examples from military science and practice.

Modularizing Risk and Plug-and-Play Processes

Maneuver-driven competition necessitates far-reaching changes in the leadership and structure of the innovation firm. For example, Nintendo did not assume that the Wii would be a runaway success when they moved into sensor based games. They also hedged their bets by (1) developing their own software for the Wiimote and (2) *not improving* the Wii's video fidelity over its predecessors so that third-party game developers could design games using traditional video controllers relatively inexpensively. In fact, it is quite possible that if Nintendo had not put these hedges in place that the Wii would have flopped. To develop a technology road map that is robust against unpredictable or unforeseeable outcomes, an innovation leader must learn how plan for contingencies. One key step in this is to *modularize risk*.

Because of the nature of technology road maps, later projects necessarily depend upon the personnel and capabilities developed during previous projects.[18] Thus, unless the risk in these individual projects is not compartmentalized, the failure or delay of one project can create a domino effect that can derail the entire road map of a firm's planned projects.[19] However, the innovation leader can intelligently insulate the dependencies that occur between dependent projects through various tactics such as providing time buffers at the end of projects that are on the critical path for many follow on projects. Such a buffer ensures that when that project is delayed, it does not spread the risk on follow-on projects. Another option for modularizing and managing risks is cross-training employees for rapid redeployment and utilizing knowledge management systems for rapid reutilization of previously developed knowledge. Another crucial way to modularize risk occurs during the roadmapping phase of the SROM cycle, when scenario planning is used to identify key capabilities that employees might need under all plausible scenarios and develop them in-house, *even if those capabilities are not necessary for the most likely scenario*. This is akin to boy and girl scouts, who carry their rain gear when backpacking even on the sunniest day. Other techniques for modularizing risk can be employed in the innovation system. One is to engage in a process of classifying capabilities with an ABC scheme similar to that used in inventory management.[20] Under this scheme,

---

[18]Fleming and Sorenson describe foundational issues associated with technological searches. Loch et al. discuss the risk management implications of such searches during the management of innovation projects. We build on these ideas by pointing out that path dependence, along with market feedback, alters the underlying search landscape across generations of development projects.

- Fleming, L., Sorenson, O.: Science as a map in technological search. Strategic Manage. J. **25**(8–9), 909–928 (2004).
- Loch, C., DeMeyer, A., Pich, M.: Managing the Unknown: A New Approach to Managing High Uncertainty and Risk in Projects. Wiley (2006).

[19]The idea of modular actions as a basis for maneuver-driven competition draws upon the work of Clark and Baldwin:

- Clark, K.B., Baldwin, C.: Design Rules. Vol. 1: The Power of Modularity. MIT Press (2000).

[20]The exact origin of the ABC technique is unknown. See Stevenson, W., Hojati, M.: Operations Management: McGraw-Hill/Irwin (2005), for a description.

"A" capabilities must be developed in full; "B" capabilities must be developed only as minimal kernels upon which to expedite should they be needed, and "C" technologies can be left to outside suppliers. This idea and several others are described in Chap. 6: Modularizing Portfolio Risk, along with an example drawn from an innovation project in the automobile industry. Another enabler of maneuver-driven competition is the adoption of *plug-and-play processes and procedures* by all innovation teams in a firm. For example, when Nintendo realized that it was reaching the female demographic with the Wii, managers shifted their game developers from the latest installment of *The Legend of Zelda*, a relatively traditional high-fantasy videogame, to developing the Wii Fit, not even a videogame in the traditional sense. If an innovation firm can move their capabilities around in a plug-and-play manner because of standardized business processes and procedures, this simplifies the SROM planning cycle in several ways. One is that it expedites the scouting process by simplifying information scaling because innovation leaders do not have to determine what a subordinate innovation manager, outsourcing partner, or allied open-source community is likely to do in a given situation. Instead, with the standardization of plug-and-play procedures, innovation leaders already know how their subordinates will respond. For similar reasons, it makes roadmapping simpler because the reaction of all subordinates becomes more predictable, thus limiting the effect of unforeseen innovation butterflies. Implementation of plug-and-play processes also expedites the orchestration phase because only general goals and intent, rather than detailed instructions, need to be communicated to subordinates. Lastly, the maneuver phase is likely to speed up as well because, with plug-and-play processes, less time will be spent on reinventing the wheel. In addition to speeding up the SROM cycle, plug-and-play processes also make it more flexible. For example, if personnel need to be transferred from project to project under a contingency plan, they should become productive much more quickly because the processes and procedures should differ little between projects. These ideas are further discussed in Chap. 6: Plug-and-Play Capabilities with an example from MedDev Inc., a leader in the medical devices industry.

Innovation Leadership and Culture

In the third section of the book, we turn to the nature of the innovation leader's work, which to a great extent, is to create and nurture a culture in which maneuver-driven competition can flourish while suppressing the worst effects of decentralization generated by the self-interest of individuals within the innovation firm. While we know that Nintendo has a somewhat unique culture that encourages innovation (e.g., design guru Shigeru Miyamoto has shown a willingness to change design late in the development cycles),[21] we are not familiar enough with the firm to actually

---

[21] http://www.wired.com/gamelife/2009/06/shigeru-miyamoto-interview/.

describe how their leadership accomplishes this or whether such techniques would play out in a non-Japanese setting. In fact, how to create and sustain an innovative business culture is not well understood in general. Intriguingly, Peter Senge when discussing the creation of a learning culture in *The Fifth Discipline* asked, "Who is the most powerful person in a ship? The ship's captain or its architect?" We argue that in an innovation culture, the choice is not binary. An innovation leader must not only be the architect, and the captain, but also the firm's coach. The rationale behind our argument and each of these roles—architect, captain, and coach—are further discussed in Chaps. 8–10 with the help of case studies of innovative leaders in the fields of architecture, global exploration, and sports.

By acting as an architect for the innovation firm in terms of capabilities and technology, the innovation leader can shape a future that harnesses the energy of the innovation butterfly while minimizing the dangers of a potentially raging tornado. By acting as the innovation firm's captain, the innovation leader can—much like a ship's captain—ensure that this road map is followed through the SROM cycle. However, to do this successfully requires developing a culture of leadership in which the subordinate innovation leader and her team are empowered to act on their own without constant feedback and direction from upper management. To enable this, empowerment must consist not only of the authority to make decisions without fear of micromanagement but also with the knowledge from upper echelons as to why an objective has been given, so that a subordinate leader knows what to do in case that objective is suddenly dropped or begins to conflict with another objective. Finally, the innovation leader must also be like an athletic team's coach by developing a playbook of shared business processes that enable effective action by subordinate leaders without their having to "reinvent the wheel" as circumstances change. Of equal importance, innovation leaders at all levels must also emulate coaches by cultivating a set of norms and values for the firm that stress the mutual responsibility of the firm to its innovation workers and vice versa. By doing so, many of the problems of decentralized opportunism stemming from individual self-interest can be ameliorated. To illustrate these ideas in more detail, we briefly discuss the leadership philosophies of famous examples from history that embody these ideals. For the leader as architect, we discuss the career of Isambard Kingdom Brunel, who developed the water-tight compartments used to improve safety in ocean-going ships. For the leader as ship's captain, we use the example of Captain James Cook, who not only was responsible for many firsts in his three voyages of discovery, such as the circumnavigation of Australia in 1776, but was also the first ship's captain who managed such an exploration without losing the lives of the majority of his crew, experiencing a mutiny, or both. (He was also, interestingly, the inspiration for the character of James Kirk in Star Trek.[22]) Finally, we discuss the coaching philosophy of football coach Bill Walsh, who led the San Francisco 49ers to the Super Bowl three times, and who spent much of his time cultivating an unconventional and innovative organizational culture for his team.

---

[22]Shatner, W.: Up Till Now: The Autobiography. Macmillan (2008).

In addition to these discussions, in the appendices, we present a glossary of the complexity terms we use in this book. We also present some additional tools for identifying, tracking, and leveraging data as well as scenario and simulation analysis. Some of these tools come from disparate disciplines of complexity science, systems engineering, and systems thinking. Yet all of them can be useful during planning and oversight of complex innovation systems.

The Risk of the Innovation Butterfly

The risks that arise from innovation butterflies cannot be overstated. As mentioned previously, from surveys of our students, many of whom represent firms well known for innovation, we have found that their teams are spending at least one-third of their time coping with the effects of the innovation butterfly. This is a significant share of an innovation team's resources, especially during difficult economic conditions. In some settings, emergent effects can explode out of control and rip any possibility for shaping the future out of the hands of the individual project manager even before the project is launched because of cost overruns or lack of competitive positioning.[23] Sooner or later, an innovation butterfly will lead to a tsunami of change in most industries. Just as the evidence from the videogame case suggests, it is an unfortunate yet simple fact that most innovation firms eventually succumb to these tsunamis. The question then for all innovation leaders is whether they want their firms to survive during "their watch." We, therefore, present the concepts and tools over the next ten chapters to empower the innovation firm not only to survive, but also to thrive in a world of innovation butterflies.

In the next chapter, we explain in more detail what creates innovation butterflies, how the innovation system amplifies their effects, and how the innovation leader might track and forecast their behavior.

---

[23]Colleagues David Ford and Tim Taylor have studied situations where innovation processes cross over a tipping point that separates success from failure. They use arguments that are analogous to our discussions in part I to describe—(a) how either external forces or internal structure can push a project out of control and to failure and (b) how the delayed and nonlinear impacts of a corrected external problem can cause a project's internal dynamics to evolve from success into failure.

- Ford, D.N., Sterman, J.D.: Overcoming the 90% syndrome: iteration management in concurrent development projects. Concurr. Eng. Res. Appl. **11**(3), 177–186 (2003).
- Taylor, T., Ford, D.N.: Tipping point failure and robustness in single development projects. Syst. Dyn. Rev. **22**(1), 51–71 (2006).
- Taylor, T., Ford, D.N.: Managing tipping point dynamics in complex construction projects. ASCE J. Constr. Eng. Manage. **134**(6), 421–431 (2008).

Part I
Understanding the Nature
of Innovation Butterfly

It has been said: The whole is more than the sum of its parts. It is more correct to say that the whole is something else than the sum of its parts, because summing up is a meaningless procedure, whereas the whole-part relationship is meaningful.

Kurt Koffka, Principles of Gestalt Psychology[1]

The butterfly effect mentioned in the first chapter describes the dependence of system behavior on small changes in the initial condition. This effect can be exhibited by very simple systems: for example, a ball placed at the crest of a hill might roll into any of several valleys depending on whether it is even a tenth of a millimeter off in its initial placement. Adding complexity such as multiple hills or rolling multiple balls that might bounce against one another just makes the outcome more difficult to predict. In this type of physical system, one can examine the emergent behavior (or butterfly effect) by running numerous tests and by building elaborate mathematical models. In social systems, this is clearly impractical, which makes the innovation leader's job much more difficult. However, by understanding the structure of the innovation system better, the innovation leader can improve her ability to plan and react to innovation butterflies. Developing this understanding is the goal of the next three chapters.

Recall that we use the terms "butterfly" effect and emergence synonymously. Emergent phenomena have been studied by complexity theorists, typically within the context of physical systems. For instance, Yaneer Bar-Yam's text *Dynamics of Complex Systems (1997)* offers empirical observations and many mathematical models to examine a variety of complex systems ranging from neural networks that model human brain, Darwinian fitness functions used to study the evolution of

[1] Koffka, K.: Principles of Gestalt Psychology. Harcourt-Brace, New York. p 176 (1935).

organisms, to computer models that analyze the implications of social policy. Within physical systems, complexity is typically studied by observing *both* the individual components and their collective behavior. For example, gas particles follow laws of statistical mechanics in terms of their individual position and velocity, but their emergent behavior becomes transparent only when we have many particles such that collective gas dynamics can be described in terms of aggregate properties such as air pressure and temperature. The study of complexity physics has advanced to the extent that now there are laws described in terms of mathematical models that can actually guide our understanding of the emergent behavior of physical systems.[2]

Some readers might then wonder if complexity studies in the social sciences have progressed to the extent that innovation systems can be analyzed to understand the behavior of the innovation butterfly.[3] That is, can the interaction between market expectations and product performance in the video game industry that was described in the previous section be understood by adapting the mathematical principles of complex systems to analyze market research data and the evolution of technical feature sets? The answer, based on the current state-of-the-art, is that innovation systems have certain unique features that make them different from systems that strictly follow only the laws of physics, making it difficult to define mathematical laws that describe their emergent behavior.

The key feature that sets innovation systems apart from purely physical systems is the involvement of human actions and their decision-making. These actions interact with the structure of the system, e.g., who shares information about innovation project progress with whom, and how often does this sharing take place. Even well-informed managers can easily err in judgment coordinating such a system owing to delays in receiving information, inexact understanding of consumer preferences, and biases in how they make their decisions.[4]

Because of these problems, we will not even attempt to present any mathematical models in this book. Instead, we will describe innovation system behavior in terms

---

[2]Examples of such models within complexity science include the iterative maps that can yield limit cycles and chaotic outcomes, phase transitions in thermodynamic systems, the rule of cellular automata, and fractals, scaling and renormalization.

Bar-Yam, Y.: Dynamics of Complex Systems. Perseus, Cambridge, MA (1997).

[3]An emerging set of papers describe the applicability of complexity theory as it relates to engineered systems. See, for instance:

Braha, D., Bar-Yam, Y.: The statistical mechanics of complex product development: empirical and analytical results. Manag. Sci. **53**(7), 1127–1145 (2007).

Braha, D., Bar-Yam, Y.: Topology of large-scale engineering problem-solving networks. Phys. Rev. E **69**, 016113-2-7 (2004)

Clark, K.B., Baldwin, C.: Design Rules. Vol. 1. The Power of Modularity. MIT Press (2000).

[4]Sterman, J.: Modeling managerial behavior: misperceptions of feedback in a dynamic decision making. Manag. Sci. (1989).

of three simplified principles: *the principle of escalation of expectations, the principle of exchange, and the principle of providential behavior.*[5] The origins for these three principles draw upon parallels from physical systems. For readers who are well versed in complexity theory, we note that the *escalation of expectations* have many analogs in the study of complexity physics, e.g., time-scale dependence and the study of kinetic pathways during protein folding. In a similar manner, *exchange* can be thought of either as phase transitions in a two-state system or as attractor behavior in a network. The third principle, termed the principle of providential behavior, draws upon the concepts of self-organization, particularly when actors are endowed with similar resources and skills but somewhat different motivations.

Before going any farther, we must also note that innovation systems are becoming information rich. Many distributed development practices, such as off-shoring, outsourcing, or open sourcing, only work because of improved information and collaboration technologies in which data has become the currency of exchange. One can apply these three principles of the innovation system toward visualizing, handing, interpreting, and mining these data, but we defer the discussion of such issues to Part II and to the appendix on information analytics.

In the next three chapters of this section, i.e., Chaps. 2–4, we describe these principles and offer empirical evidence from many different innovation settings to identify the most important characteristics of the managerial decisions in the innovation system.

---

[5]Caveat: Absent precise mathematical definitions and rigorous tests, these are mere hypotheses. We take the liberty of terming these hypotheses as principles and offer evidence from multiple settings that point to their applicability. We hope that our work will spark academic debate and rigorous hypotheses testing.

It is possible to examine the impact of these principles while modeling their collective behavior by using simulation techniques. In the appendix, we provide references to simulation methodology that capture the interactions among various components that balance a product portfolio. Even a slight change, either due to an exogenous shock or a firm's decisions, can tip this balance.

Chapter 2
Escalation of Expectations over Past Performance

Complexity Arises When You Try to Please Customers

The past is never dead; in fact, it's not even past.

William Faulkner[1]

The principle of *escalation of expectations* describes the idea that customers value innovation to the extent that it surpasses their performance benchmarks, which were in turn created by the accumulation of past innovations. However, this is not a simple linear relationship in which the outcome is proportionate to the input effort. Instead, the result of innovation firm's efforts is quite nonlinear, often behaving much like the proverbial single straw that breaks the camel's back. The *escalation of expectations* is the mechanism that allows small changes, for instance, a small request made by a random customer demographic, the passage of a new law, or perhaps the wishes of a CEO, to become astronomically amplified in its impact throughout the innovation system. The escalation of expectations sets entire firms along a particular path of innovation, in which targets get set, ideas emerge, alternatives are searched for, and chances are taken. Out of this process, firms build up deeper reserves of know-how, and customer expectations ratchet forever upward (and sometimes, as we will see, even sideways).[2]

---

[1] Requiem for a Nun, Random House (1951).

[2] Dorothy Leonard (1995) was among the first scholars to offer arguments for building capability through a self-reinforcing or "virtuous" loop. She has also pointed out that *innovation capabilities* are "deeply rooted in the past," and often grow organically. These arguments have shaped our thinking about the principle of escalation of expectations in the sense that we integrate the issues of evolutionary complexity and selection through market feedback mechanisms into the broader literature on product innovation capability. This literature has evolved considerably over the past 20 years; see, e.g., an edited volume by Garud and Karnoe (2001), and a recent exposition of these ideas at Infosys Technologies by Garud et al. (2006).

E.G. Anderson and N.R. Joglekar, *The Innovation Butterfly*,
Understanding Complex Systems, DOI 10.1007/978-1-4614-3131-2_2,
© NECSI Cambridge/Massachusetts 2012

In order to explore the nature of escalation of expectations, we consider the development of the automotive industry during the last quarter of the twentieth century. The 1970 U.S. Clean Air Act and its various amendments combined with rising gasoline prices forced automotive engineers to reduce air polluting auto emissions and increase fuel efficiency. Up to that time, automotive engines were controlled primarily by mechanical means such as centrifugal spark advances and carburetors. However, the prevalent mechanical control technologies could not attain the degree of precision necessary to meet the legislative requirements associated with increased fuel efficiency. To solve this problem, the "Big Three" automotive companies in the USA shifted their engine control architecture from mechanical to electronic control. Therefore, these automakers were able to implement sophisticated software algorithms instead of mechanical manipulation to control the flow and combustion of gasoline in the engine. Electrical engineers with an expertise in control systems needed to implement this innovation stream simply did not exist within automotive design teams prior to 1970. To train and develop these engineers in sufficient numbers took automakers more than a decade. However, by the end of the 1980s, legions of them were employed by all the major automotive companies. Nowadays, engines are routinely controlled by electronics within small microprocessor devices, similar to those devices that run the personal computers.

In summary, for automotive manufacturers, a disruption created by Clean Air Act and the rising gasoline prices in the early 1970s led to a need for greater capability in electronic control system design. Firms could only accomplish this by developing sufficient numbers of in-house engineers with the appropriate training in electronics and experience in automobiles over a 10-year period. At first this capability was used simply to meet the requirements of the Clean Air Act. After raising product performance back to acceptable levels by developing a capability in electronic control systems, the automotive industry began to look for other market needs that these capabilities could fulfill. As stated by Jerry Rivard, former vehicle controls guru of Ford Motor Company:

> As integrated circuit technology evolved, it became possible to design many functions into integrated circuits, thus eliminating a lot of discrete components ... electronic engine controls were representative of how the [automotive] industry evolved vehicle subsystems.[3]

- Leonard-Barton, D.: Wellsprings of Knowledge: Building and Sustaining the Sources of Innovation. Harvard Business School Press (1995).
- Garud, R., Karnoe, P.: Path Dependence and Creation. Lawrence Erlbaum (2001).
- Garud, R., Kumaraswamy, A., Sambamurthy, V.: Emergent by Design: Performance and Transformation at Infosys Technologies. Organ. Sci. **17**(2), 277–286 (2006).

As pointed out in the body of the chapter, this logic does not apply to competence destroying innovations:

- Henderson, R., Clark, K.: Architectural Innovation: The reconfiguration of existing product technologies and the failure of established firms. Administrative Science Quarterly (1990).

[3]Rivard's views are described in www.sae.org/automag/electronics/09-2002. Many of the details in this part are based on the first authors work at Ford. For detailed discussion of the underlying case,

That is, once the electronic design capability was developed to address legislative requirements, U.S. automotive manufacturers found that they had acquired product architecture, control and software engineering capabilities that allowed them to develop a number of features that were inconceivable prior to the introduction of electronic controls. According to Rivard, this change enabled the development of such customer-pleasing features as antilock brakes, traction control, all-wheel drive, advanced maintenance diagnostics, communication and navigation systems, and thermostat-controlled air-conditioning, which had nothing to do with the 1970 Clean Air Act Requirements that created the auto industry's electronics controls capability in the first place. Interestingly, these new features were initially positioned by automotive marketers as exciting novelties, but, as in the case of antilock brakes, some of them shaped consumer preferences to such an extent that they soon became "standard" options without which a new automotive model could not compete in the marketplace.

Other features such as four-wheel steering did not find any customers and disappeared seemingly without a trace. A similar innovation–change cycle may be underway again with the advent of hybrid and electric vehicles. Some of the underlying technologies that might be ushered in by these changes involve energy storage and charging of batteries. Interestingly, one could argue that this current development, involving hybrid vehicles and perhaps remotely controlled vehicles, would have been impossible without the prior development of electronic control capabilities driven by the Clean Air legislation and allied amendments. If one compares the dimensions on which rating firms, such as J.D. Power Associates, review the performance of automobiles, it is easy to see that electronics and smart/clean technology-based performance measures are increasingly evolving into the key basis of comparison among automotive consumers. Some of the follow-on innovations like GM's OnStar system were embraced; others, such as four-wheel steering were not. In both cases, however, these developments were unanticipated results of the 1970 Clean Air Act that disrupted the industry and led to an unforeseen series of customer-pleasing innovations.

It is difficult, but possible, to study the emergence of software and electronics development capability within the automotive "ecosystem." Such studies may specify individual elements, such as the performance specifications for microprocessors or battery technologies in terms of simple sets of mathematical rules. One can then abstract these individual actions and connect them appropriately through feedback loops to explore the complex interactions that create the patterns of their collective behavior. Next, we describe a graphical methodology that will allow the reader to follow these connections we have identified between market needs and product performance within an innovation system. Later, we show that this system follows the *principle of escalation of expectations*, which reflects the collective evolution of the

---

see Anderson and Joglekar 2005 (Anderson, E.G. Jr, Joglekar, N.R.: A hierarchical product development planning framework. Prod. Oper. Manage. (2005)). The integration of mechanical and electronic technologies that is needed to implement offerings such as OnStar is discussed in Joglekar and Rosenthal (2003).

• Joglekar, N., Rosenthal, S.: Coordination of design supply chains for bundling physical and software products. J. Prod. Innovat. Manage. (2003).

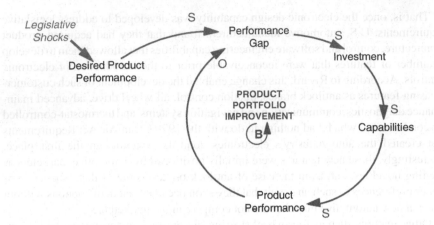

Fig. 2.1 Capability development—a balancing loop

market landscape. These behaviors can be studied more deeply with the aid of a computer simulation, as discussed in the appendix.

Our discussion of *escalation of expectations*, however, cannot complete until we discuss disruptions. From time to time, a novel type of product emerges based on a new architecture that makes the existing stock of painfully accumulated capabilities irrelevant. For instance, Henderson and Clark (1990) offer a discussion of competence destroying innovation sequences from the photo-lithographic industry. Such a disruption did in fact happen to the mechanical control engineers who designed carburetors and spark plug advance mechanisms in the automotive industry, as described earlier. Hence, the escalation of expectations results in the innovation system behaving predictably some of the time and almost randomly at other times. This is why we mentioned in our earlier discussion that market expectations can sometimes move sideways. Moreover, shifts between predictability and unpredict-ability are often unexpected. Thus, in some sense, the role of time in the *escalation of expectations* must be reset, for example, a firm's capability levels may be reduced or even wiped out and some new types of capabilities may need to take their place. Thus, understanding the impact of the butterfly effect becomes, if anything, more significant in such settings.

Evolution of Complexity

Describing the dynamic complexity of the innovation butterfly and its implications for the innovation systems with mere words is difficult. System dynamics is a social science methodology that explores dynamic complexity in industrial landscapes and offers a visual language for helping us understand them called causal-loop

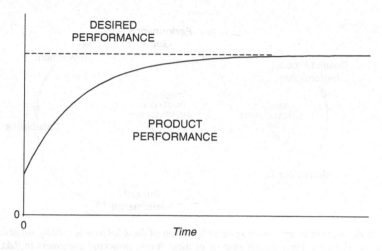

Fig. 2.2 Balancing loop behavior of product performance

diagrams.[4] A causal-loop diagram of electronics capability development in the automotive industry is shown in Fig. 2.1.

Causal-Loop Notation: S stands for support (or positive correlation), e.g., increase in desired performance increases the performance gap. O stands for oppose (or negative correlation), e.g., increasing product performance reduces the performance gap. If there is no symbol shown for ease of exposition, the link would implicitly indicate "S" relation in reading the diagram.

In Fig. 2.1, start with *Legislative Shocks*; in the language of causal-loop diagrams, this is a variable, meaning that it has a value that can increase or decrease. Because the number of legislative shocks affects *desired product performance* in terms of acceleration and other measures of engine quality, an arrow shows the causal relationship between the two variables. All other things being equal, increasing requirements, e.g., legislated standards for fuel efficiency, increases desired product performance; likewise decreasing legislative requirements decreases the desired product performance.

Following the chain of variables around the loop in Fig. 2.1, as auto engine performance improves, the gap between desired performance and actual performance decreases. An "O" at the head of the arrow links these two variables to show this opposite relationship. This in turn will reduce the need for further *investment* in that particular capability. Because *investment* moves in the same direction as the performance gap, this link is left unmarked. If the performance improves and the performance gap reduces over time, it will reduce the need for future investment in that capability. When this cycle repeats over multiple generations of product introduction, any change in capability—or any other variable in the loop—will eventually feedback on itself. When this occurs, the circular chain of linked variables is termed a "causal loop," or more simply, just a "loop." Interestingly, increasing any

---

[4]Sterman, J.D.: Business dynamics: systems thinking and modeling for a complex world. Irwin/McGraw-Hill, Boston (2000).

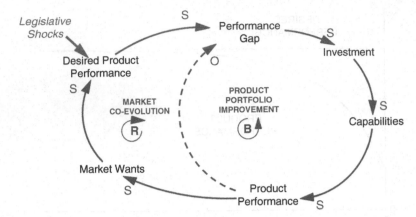

Fig. 2.3 Twin loops in an innovation system (A version of these loops was initially suggested by Professor Jay Forrester. For a formal analysis of these loops, involving investment in R&D and production capabilities, using a data set on several generations of Intel's microprocessor products, and allied market demand, see Joglekar (1997)[5])

variable in this capability loop will eventually result in a countervailing pressure to reduce further changes in that variable. These are called "balancing" loops and are denoted with a "B" within a circular arrow at the center of Fig. 2.1. The resultant evolution of product performance is shown in Fig. 2.2 in which a balancing loop creates a drive in the system toward a desired performance goal.

However, Fig. 2.2 does not fully capture what happened as the consumers came to embrace various manifestations of the electronic controls revolution in the automobile industry. The desire for product performance is never likely to remain static—it evolves over time. For instance, product features such as antilock brakes became standard equipment because of shifts in consumer expectations and desires. This second dimension to the growth of electronic control system capabilities is shown in Fig. 2.3 in the "market co-evolution loop."

The "desired product performance," term captures the formulation and escalation of the consumers' expectations for the product. As the available product performance increases, the consumer want even a better product and raise their expectations, i.e., raise the desired product performance level.

If one tracks the outer loop in Fig. 2.3, it is evident that as *product performance* improves, it prompts *market wants,* i.e., consumers' desire enhanced product performance, to also increase.

This phenomenon drove U.S. auto manufacturers to raise their *desired product performance* in order to remain competitive, thereby increasing the *gap* between desired and actual performance. This gap caused firms to increase *investment* in the capability to further develop this aspect of product performance. Over time, the

[5] Joglekar, N.: The *Technology Treadmill*: Managing Product Performance and Production Ramp Up in Fast-Paced Industries. MIT Sloan School of Management, Thesis (1997).

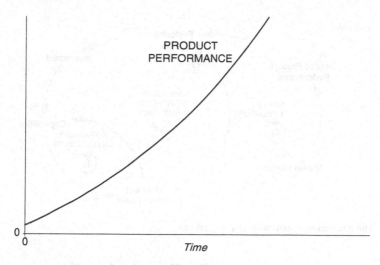

Fig. 2.4 The reinforcing effect of the market co-evolution loop

capability itself improved, resulting in improved product performance. These sorts of loop dependencies drive performance escalation as shown in Fig. 2.4. The fact that any increase in product performance (or any other variable in the loop) will result in further increases in that same variable lead labeling the loop a "reinforcing loop" and is noted in the left-hand side of Fig. 2.3 by an "R."

For convenience, this loop is identified in Fig. 2.3 as the "market co-evolution loop," because product performance, firm capabilities, and market desires co-evolve and lead to a reinforcing effect between the variables involved in this loop. However, such a reinforcing behavior may operate both ways. If a product's performance declines, eventually consumers will adjust their behavior and demand for this product will decline—although this may take a long time—changing what had been a "virtuous" cycle of growth into a "vicious" cycle of declining demand. For instance, with the advent of sensor technologies in videogaming, the competitive focus and market demand have moved away from improving the graphics and rendering quality, as discussed for the Wii case in the previous section.

Embedded Complexity[6]

"Their work was, as it were, a wheel within a wheel..." Ezekiel 1:16

This is far from the end of the loops within loops that complicate the management of innovation systems. Typically, another reinforcing loop is also at play. The practice of any given capability will ultimately result in individual and organizational

---

[6]Complex physical systems are endowed with nested or replicated patterns connections (see for instance, http://en.wikipedia.org/wiki/Mandelbrot_set) for a discussion of Mandelbrot Set.

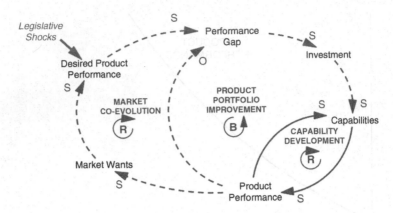

Fig. 2.5 The innovation system with learning effects

learning, leading to an improved capability. This is the classic "learning curve" as it exists in the product development world. It is represented by the "Capability Development" loop shown in Fig. 2.5.[7]

However, we have not yet fully examined how the escalation of expectations (i.e., how the effect of time impacts the outcome) plays out in the product innovation system. In particular, we have not yet considered the fact that variables such as desired product performance, investment and the realized product performance have delays between their cause and effect. For instance, the investment decisions may be a part of an annual planning cycle, while the product performance will only be visible to the consumers after each new generation is launched, which might in some industries occur only every few years. In Fig. 2.6, the most significant delays are marked.[8]

In most industries, the delay in forming market wants is relatively long compared with those of the other two delays, which are associated with product portfolio improvements. However, these long delays are crucial to understanding the system because even managing a very simple feedback loop with a long delay can

Such nested patterns also occur within an innovation system at lower levels of abstraction: a project, a task or subtask, and so on (see Sosa et al. 2007). For ease of discussion, we exclude nested loops that occur in lower levels of abstraction.

- Sosa, M., Eppinger, S., Rowles, C. A Network approach to define modularity of components in complex products. J. Mech. Des. (2007).

For discussion of tipping points and disruptive innovations, see:

- Gladwell, M.: The Tipping Point: How Little Things Can Make a Big Difference. Little Brown, Boston (2000).
- Christensen, C.: The Innovator's Dilemma: When New Technologies Cause Great Firms to Fail. Harvard Business School Press (1997).

[7]This graphic is a simplification of the reality. For the learning to accrue, teams must be incented, coached, enabled and their success be celebrated, such that such effort becomes a part of the organizational culture. Some of these issues are addressed in Parts II and III.

[8]The effects of delays can be particularly insidious in managing the innovation systems. David Ford and John Sterman have modeled many aspects of their effects in the presence of rework.

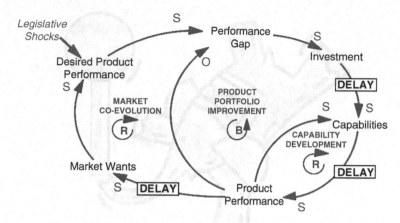

Fig. 2.6 The innovation system including delays

be difficult. As a very simple example, consider your morning shower. When you step into a shower and it is too cold, the natural reaction is to crank up the hot water. Usually, however, the water does not heat up for at least 30 seconds. Because we are often impatient, we crank up the hot water handle further. Then, finally the hot water begins to flow through the pipes to the faucet, at which time we generally discover that we have made the water far too hot and we jump away from the scalding shower. Oftentimes, this is immediately followed by turning the hot water handle down too much, followed by the inevitable cold shower.

This shower example contains only a single loop in terms of the action and reaction between the person controlling the shower temperature and the desired temperature. It also features a delay between the person raising temperature and hot water flowing through the faucet. Similarly, in the innovation system's diagram shown in Fig. 2.6, we have three separate delays, each of which is of a different scale that ranges from weeks to months or even years. One can imagine that managing this system of delays is an inherently challenging task. And, in fact, this is true. Numerous studies have shown that our trouble in adjusting the shower temperature also appears when controlling management systems, only worse (for a survey of the literature on managing feedback and delays, see Sterman 2000). The inherent difficulty in managing the underlying complexity within these three loops is somewhat akin to the act of an elephant balancing a beach ball on top of a long pole, which is in turn balanced on tip of its trunk, as shown in Fig. 2.7.

The elephant has to manage a dynamic system similar to the shower example, but this elephant has to worry about multiple points of balance rather than one. It also has to account for something mathematicians call nonlinearity, a mathematical term

If one considers the work flowing in their system as a proxy for innovation that is subject to hidden rework that surfaces (e.g., unseen customer needs), when the project progresses then their model examines issues such as how concurrence and delays will affect overall progress.

- Ford, D.N., Sterman, J.D.: Overcoming the 90% syndrome: iteration management in concurrent development projects. Concurr. Eng. Res. Appl. **11**(3):177–186 (2003).

Fig. 2.7 Managing a nonlinear dynamic system[9]

for the uneven relationship between cause and effect. In other words, unlike the shower example, where how much you turn the hot and cold handles roughly corresponds to a proportionate—and hence linear—change in temperature, the elephant would face a tricky situation, even if the stick moved an inch or two. It needs to exert much more than double the effort to adjust a ball that is only two inches off its balance point rather than one that is only an inch off because this elephant has to handle a number of different nexuses of balance that influence the other. In other words, the force between the ball and the stick has a nonlinear connection. The nonlinearity between cause and effect will increase to the point that the elephant will have to go through some wild, if humorous, gyrations to recover. (Any readers who have been to the circus can appreciate this.) Similarly, anyone who has learned to ride a unicycle while juggling simultaneously will have experienced similar nonlinear effects. It can be done, but it requires practice, and most first timers fall many times and need many do-overs before they get the hang of it.

Getting back to the innovation system loops—they too are nonlinear and similarly difficult to control. To make things worse, because each innovation is somewhat unique, managers rarely are permitted the luxury of do-overs. These sorts of systems, which are characterized by multiple feedback loops with embedded delays and nonlinear relationships between cause and effect, are referred to in physics as "dynamically complex" systems. This is the source of the nonintuitive and difficult-to-manage behaviors in the innovation system. As we mentioned earlier, these

---

[9] This drawing was made with Microsoft PowerPoint and is used with the permission of Microsoft, Inc.

nonlinear feedbacks may even drive the behavior of the innovation system to appear random in some situations, which is what a mathematician would term "chaotic." Strictly speaking, chaotic behavior is not truly random, it only appears so an observer. However, chaotic processes imitate random behavior closely enough to often frustrate innovation planners.

The good news is that the planning, development, and ultimate acceptance of innovations like antilock brakes are *not* chaotic processes. Ultimately, they will eventually reach some sort of equilibrium level of acceptance. However, the bad news is that managing innovations is still far from simple, being much like the elephant's balancing trick. Worse, they are extremely sensitive to initial conditions. For instance, if the original 1970 Clean Air Act had been just a bit less stringent or microprocessor technology had not been available in the late 1970s, it is quite possible that automotive companies may have gone to variable-venturi carburetors (an alternative mechanical technology that could improve emissions), which would have delayed or perhaps blocked the development of electronic capabilities that resulted in outcomes such as antilock brakes and the evolution of OnStar.

Thus, only a slight difference in initial conditions can evolve into at least two, and more likely several, radically different ultimate results. As evidenced by this case, emergence resulting from the competition among multiple alternatives is the most common behaviors of complex systems seen in the management arena. Malcolm Gladwell describes emergence in this context using the language of "tipping points," the point beyond which a potential emergent path metamorphoses from a possibility into inevitability. For example, the fact that drivers drive on the right-hand side of the road in the USA rather than the left side (as in Great Britain) is an emergent phenomenon that results from a tipping point being reached. Another term used by business *scholars* to describe such nonintuitive emergent phenomena is "disruptions" because they cause one apparently stable business system to rapidly evolve into something completely different. Because these disruptions in business systems, often caused by innovations, typically result in the wholesale failure of many leading firms and sometimes even industries, the great economist Joseph Schumpeter referred to the process of emergence as "creative destruction."[10]

However, the difficulties described so far in managing the escalation of expectations in product innovation are far from complete. To begin to fully comprehend the nature of those portions of the system that can lead to tipping points and business disruptions, we need to consider the effects of random shocks and other uncertainties in the system as well: in particular, what is the source of innovation butterflies?

Randomness and variability impact the system as shown in Fig. 2.8 at a number of points in the innovation system (so much so, in fact, that some experts have

---

[10]Schumpeter, J.A.: The Process of Creative Destruction. Capitalism, Socialism and Democracy, pp. 82–85 (1942).

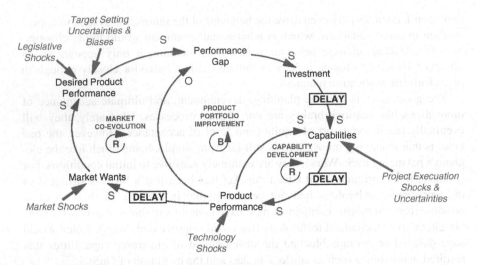

Fig. 2.8 The innovation system with multiple uncertainties

argued that elemental (or component) uncertainty is the central driver of complexity in an innovation system[11]). The effect of *legislative shocks* has been described. Unexpected advances in or negative effects of technology can also impact product performance. This effect may have either a positive or a negative direction, however, because it is unclear in which direction it will drive product performance. The direction depends on the context. For example, improved technology generally results in improved product performance, *if the technology works as expected.* If for some reason a new technology has some unforeseen drawback(s) or negative outcomes, it could actually drive product performance down. An excellent example of this phenomenon is the tendency of plasma screen television displays pixel clarity to deteriorate within 3–4 years, which was not the case with traditional televisions. This opened the door to competing technologies such as the DLP screens.[12] Similarly, market shocks may drive market desires in sudden unforeseen directions, particularly if an unexpected esthetic or fad arises. For example, the recent preference for predistressed blue jeans clearly reduces the longevity of those jeans over those made for an older generation.

Target-setting uncertainties are related but subtly different because they result from random shocks stemming from the difficulty in accurately determining market desires. Translating them into useful design specifications that are

---

[11] Suh, N.P.: Complexity: Theory and Application. Oxford University Press (2005).

[12] Digital Light Processing (DLP) is a trademark owned by Texas Instruments, representing a technology used in projectors and video projectors. The image is created by microscopically small mirrors laid out in a matrix on a semiconductor chips. www.cnet.com/1990-7874_2-5108443-3.html.

meaningful to engineers introduces yet more uncertainty into the process.[13] For instance, customer focus groups described one pickup truck produced by a U.S. firm as having "less acceleration" than its foreign competitor. The engineers assigned to develop this pickup truck could not understand this complaint at first. In the U.S. automotive industry, the standard measure to assess the acceleration has been based on the time takes to go from standing position, i.e., how many seconds it takes to go from 0 to 60 miles/hour. The designed vehicle was superior in all the standard tests of acceleration. However, deeper probing of the focus groups revealed that customers were more interested in accelerating quickly, while already in motion, in order to pass other vehicles than in reaching 60 miles/hours quickly from a standing start. This would require designing and testing the vehicle for different standards than 0–60. Additionally, it turned out that the same exact acceleration pushed people more deeply into the seats of the foreign designed pickup than into those of the U.S. vehicle. Hence, customers perceived—incorrectly—that they were accelerating more quickly in the foreign pickup.

Finally, execution shocks can also affect individual projects in a complex manner. For example, many of the worries surrounding the late delivery of the Boeing 787 Dreamliner, revolve around the impact that delay will have upon Boeing's other projects. In particular, the diversion of engineering resources to cope with the 787's delays is blamed for allowing "more engineering errors [to escape] than what would be considered normal" during the development of the new 747-8 (a modernization of the venerable Jumbo Jet),[14] thus creating a domino effect of delays in one project begetting delays in subsequent projects.[15] Moreover, as we see in the next chapter, execution shocks can result in even more complex chains of consequences.

We have described a multiple set of shocks that feature legislative, target setting, market, technology, and execution uncertainties. In the end, the net effect of these multiple sets of uncertainties is to render the management of a complex dynamic system underlying product development extremely difficult *because any one of them is a potential innovation butterfly.* Behavioral studies on people managing dynamically complex innovation systems uncertain input are scarce, but the

---

[13] For a treatment of the translation process see Griffin and Hauser (1993) and von Hippel (1988). For a discussion of the randomness associated with such processes, see Khoo and Ho (1996).

- Von Hippel, E.: The Sources of Innovation. Oxford University Press, New York (1988).
- Griffin, A., Hauser, J.R.: The Voice of the Customer. Market. Sci. (1993).
- Khoo, L., Ho, N.: Framework of a fuzzy quality function deployment system. Int. J. Prod. Res. (1996).

[14] Cohen, A.: "Boeing 747-8 Delay to Death by a 1000 Cuts," Seattle.pi Blogs, posted on 8 October 2009. http://blog.seattlepi.com/aerospace/archives/181424.asp (2009). Accessed 27 May 2010.

[15] Gates, D.: "747-8 Delay Causes Doubts at Boeing." The Seattle Times, 7 October 2009. http://seattletimes.nwsource.com/html/boeingaerospace/2010013013_boeing07.html (2009). Accessed 27 May 2010.

few studies on this indicate that random events diminish what little managerial capability exists to manage them.[16]

As the examples in this chapter show, it is difficult to capture the scope of the time dependence and interconnectedness of any system. Unsurprisingly, managing risk in even a simplified product development system characterized by dynamic complexity and randomness in an optimal manner surpasses the cognitive capabilities of managers, even when they are given high levels of computer support.[17] This should not be surprising. Borrowing from our balancing elephant metaphor, the elephant has trouble enough balancing just one ball on a flat surface at a circus arena. Staging this balancing act on a bumpy lawn outdoors on a windy day with gusts striking the ball from all directions is unlikely to improve the elephant's ability to keep the ball on its trunk and off the ground. Innovation leaders face exactly the same problem, which is precisely why managing the innovation butterfly is so difficult.

While the principle of the escalation of expectations has broad implications for planning and managing innovation, the examples in this chapter actually understate some of the other issues associated with managing complexity in the innovation system as we see in the next two chapters. So what is an innovation leader to do? A number of potential solutions exist for managing the escalation of expectations that can keep an innovation butterfly from turning into a destructive tsunami. We describe these in Chaps. 5–7. But, each solution may create another set of problems. We discuss this problem, and its resulting tradeoffs, as we turn our attention to the *principle of exchange* in the next chapter.

---

[16] We offer a detailed discussion of such biases in Chap. 4.

[17] This problem is what scientists refer to as NP-Hard, meaning that the time to solve the problem increases more rapidly than the number of different states taken to any arbitrary power (Anderson and Joglekar 2005). Practically speaking, solving any managerial problems of such difficulty in an optimal manner is essentially impossible. Some compromises must be made, such as ignoring certain feedbacks in Fig. 2.8, such that the problem can be decomposed into easy to understand pieces.

- Anderson, E.G. Jr, Joglekar, N.R.: A hierarchical product development planning framework. Prod. Oper. Manage. (2005).

Chapter 3
Principle of Exchange

All You Ever Do Is Exchange One Set of Problems for Another

> The Operative: "It's worse than you know."
> Mal: "It generally is."—Serenity, A 2005 film by Joss Whedon

In this chapter, we use a case study from the pharmaceutical industry to illustrate both the dynamic and structural complexity of the innovation system, as well as the complications arising from the fact that firms rarely work on single projects, but rather on portfolios of projects. Furthermore, firms require not just one, but rather a number of employee capabilities to deliver the portfolio to market. The multiple linkages between these capabilities, product portfolios, and the rest of the innovation system result in structural complexity that causes any "solution" to a local problem to often fail. Worse, if the local solution does work, the dynamic complexity of the innovation system means that the solution will generally create another set of problems somewhere else in the system. An illustration of this *principle of exchange* is illustrated with a description of river engineering solutions to cope with floods and other problems along the Mississippi. The partial dismantling of levees to prevent floods in urban areas is a good example of this. One area is saved from flooding, but other areas are then flooded.[1] Geoffrey Parker, Professor of Management at Tulane University, described policy fixes of complex systems in this way: "Many policies won't work, but even when they do, you simply exchange one set of problems for another. Hopefully, you end up with a better set of problems."

---

[1] It has been argued that New Orleans would not have flooded in 1927 even if a levee had not been dynamited at Caernarvon. The Caernarvon action was rendered moot because of many "natural" breaches created by the flood of 1927 upriver of New Orleans. Except where noted otherwise, all of the discussion in this chapter on the Mississippi has been drawn from Barry (1997).

- Barry, J.M.: Rising Tide: The Great Mississippi Flood of 1927 and How It Changed America. Touchstone, New York (1997).

E.G. Anderson and N.R. Joglekar, *The Innovation Butterfly*,
Understanding Complex Systems, DOI 10.1007/978-1-4614-3131-2_3,
© NECSI Cambridge/Massachusetts 2012

We generalize this observation in the Principle of Exchange: *In distributed innovation systems, management solutions to local problems, even if they are effective, may result in a set of problematic emergent outcomes elsewhere in the system.*

Structural Complexity: Connectedness

In the previous chapter, we showed that much of the difficulty in managing the innovation butterfly results from the nature of the innovation system, which is dynamically complex. However, we have if, anything, simplified the complexity of the system by ignoring the fact that there are multiple firms, each with a number of products, competing for numerous markets simultaneously. That is, up to this point, we have emphasized the dynamic complexity (number of nonlinear feedback loops) of the innovation system at the expense of ignoring its structural complexity (how many interconnections the system has). Yet both forms of complexity are important. This is true not only of the automotive industry, but also of any other innovative industries, such as pharmaceuticals.[2] For example, consider Exubera, the inhaled insulin product for diabetics, pioneered by Nektar in the late 1990s.[3] Today, diabetics must rely on several of these painful injections of insulin, which their bodies can no longer produce, to maintain a sufficient level of blood sugar throughout the day. If diabetics do not maintain this "tight" blood sugar control level, the onset of, and complications associated with, diabetes such as blindness, kidney disease, and gangrene in their extremities can occur earlier in the disease process than they would otherwise. For these reasons, according to market studies, most diabetics would prefer administering their insulin in some less intrusive manner, such as via inhalers like many asthma patients use, to control their disease. Nektar met this challenge by developing a system over a 10-year period that consisted of a delivery device that

---

[2] For a thorough coverage of the strategic dynamics in the pharmaceutical sector, see:

- Peck, C., Paich, M., Valant J.: Pharmaceutical Product Strategy: Using Dynamic Modeling for Effective Brand Planning, 2nd edn. Informa Healthcare, New York, NY (2009).

[3] Our discussion of Exubera draws upon the following sources:

- E&N, (2008). EXUBERA AND NICE, Stanford Graduate School of Business Case OIT-80.
- DOC: Weak Sales Lead to Exubera's Market Withdrawal, DOC NEWS December 2007 vol. 4, no. 12 (2007).
- WSJ: Insulin Flop Costs Pfizer $2.8 Billion, Wall Street J. http://online.wsj.com/article/SB119269071993163273.html (2007).
- Hambrecht, W.R.: Supply, Not Demand Delayed Exubera, Analyst Report (2006).
- Readers Digest: Meet Your Diabetes Support Team, available at http://www.rd.com/living-healthy/diabetes-management-and-support/article30340.html (2010).

atomized pellets of powdered insulin and suspended this powder in the delivery device chamber, from which diabetics could inhale the insulin into their lungs. However, despite the technical success of the product, Nektar did not have the capability, e.g., geographic access needed to reach physicians, to take this product to market. So it partnered with Pfizer, a major player in the pharmaceutical industry. Together, they launched a commercial version of Exubera in 2006 with a large marketing budget and much media fanfare. However, this product's low sales and slow adoption rate eventually led it to being withdrawn from the market in 2007.

At first, it was thought that the poor acceptance rate resulted from a misreading of the customers' enthusiasm for the product. However, the picture that results from a close examination of the data is somewhat more complex. Pfizer experienced delays in bringing the production of the Exubera pellets of insulin up to planned volume. So they decided not to begin an all-out marketing campaign to shape the tastes and expectations of physicians who treat diabetic patients while there was a potential shortfall of pellets to meet demand. Instead, Pfizer chose only to pursue a limited campaign targeted at doctors who specialize in diabetes management. This limited push had several negative consequences. One is that the number of patients treated by diabetes specialists is relatively small (approximately 10% of the overall market). Hence, Pfizer missed out on a chance to shape the expectations of the vast majority of the physicians, particularly with respect to the fact that marketing stud ies indicated that many more patients are willing to begin insulin treatment and properly follow it if inhalation is the mode of delivery rather than injection. This resulted in a sizable slice of the market being left on the table from which positive word of mouth could never develop.

Moreover, things did not work well with those physicians who were targeted either. Physicians who specialize in treating diabetes typically encourage patients to maintain the "tight" regime of blood sugar control with four or perhaps more injections or inhalations per day. Unfortunately, delivering the correct insulin dosage for each inhalation proved problematic for many patients. Because of the multiple dosages involved in a tight regime, lower dosages are typically administered creating a need for large numbers of the smallest Exubera dose, the 1 mg pellet. Presumably because of the production bottleneck described earlier, the 1 mg pellets were sold only in bundles with 3 mg pellets. This caused patients to purchase many unused 3 mg pellets or, alternately, not to have enough 1 mg pellets on hand, interfering with "tight" blood sugar control. Inhaling the correct dosage was made even more difficult by the nonlinear relationship between standard international units of insulin used for injection, which most diabetic patients were familiar, and dosages of Exubera. Thus, complying with a tight regime under Exubera turned out to be more difficult than anticipated for patients, thus defeating what should have been one of Exubera's greatest advantages over injected insulin.

Other issues existed as well. The inhaler was bulky and unattractive, and because of the fear of Exubera inhalation leading to lung capacity issues, the patient was further inconvenienced by needing to take extra lung capacity tests. Combined with the other issues described above, many insurers decided not to cover the 2–3 dollar difference per day in treating patients with Exubera vs.traditional injection insulin therapy.

It has been speculated that Pfizer could have avoided this last trap if it had specifically addressed the concerns of insurance providers in its development and marketing campaign. However, Pfizer did not do this. Instead, an initial execution failure in production ramp-up drove the market-co-evolution-reinforcing loop connecting the market wants of consumers, physicians, and insurance companies to investment in product performance in the wrong direction. This resulted in the market not only rejecting Exubera but also becoming wary of all other types of inhaled insulin therapies. Even if Pfizer or any of its competitors eventually develops a safe, easy-to-use inhaled insulin product without any of Exubera's issues, that product will face much more of an uphill battle in gaining acceptance because of Exubera's negative shaping of market tastes. Accordingly, most of Exubera's potential competitors have terminated their efforts to enter the market.[4]

In short, it seems that market acceptance of Pfizer's Exubera insulin inhaler was weak because they lacked the capability to ramp up their production of the Exubera medicine to satisfy potential demand. Yet, this description still does not capture the full scope and interconnectedness of the innovation system because it neglects the fact that firms generally make more than one product and serve various markets. Indeed, Pfizer has reported that it was developing a number of other products during the period when Exubera was launched. Like Boeing's issues with the 787 and the 747-8 described in the previous chapter, there could very well have been some tight interdependencies across their product line. For example, while we cannot confirm this, it is quite possible that the diversion of some of its manufacturing capabilities (such as engineering talent or actual production capacity) to a product other than Exubera was responsible for its inability to quickly ramp up for full production volume of the Exubera insulin inhaler.

This is not to imply that the linkage between products and markets has a simple one-to-one mapping. For example, Exubera could potentially have served both Type 1 and Type 2 diabetics. However, the needs of each of these consumer groups for Exubera were subtly different. Type 1 diabetics, whose disease arises from an auto-immune disease that destroys the body's ability to make insulin naturally, must externally administer synthetic insulin several times per day to control their blood sugar levels. However many Type 2 diabetes patients, whose bodies can still produce some insulin naturally, do not need to administer synthetic insulin as often as Type 1 diabetics, so the attractiveness of abandoning injected insulin is, relatively speaking, lower. There exists difference even among different groups of Type 2 diabetics. For example, some Type 2 diabetics do not require insulin injections and can instead take insulin in the form of an oral pill. Others do not require insulin and may be prescribed alternative treatments. Type 1 diabetics also have large subgroups that would not benefit from inhaled insulin, such as those who also suffer from asthma. Thus, multiple, dynamically complex linkages exists between the development of various products a company offers and the evolution of markets it serves.

---

[4] From In-PharmaTechnologist.com: Lewcock, A.: "Novo drops inhaled insulin plans in post-Exubera fallout," January 15, 2008.

Because the performance of each product is connected not only to its various markets, but also through these markets, connected to other products, the linkage between various products forms an intricate network of connections. Importantly, this implies that network relationships exist between the various markets that a firm services, and also between the desired performance targets for each and every product along multiple dimensions. Many authors have termed all of a firm's products, when considered as a whole, as the firm's "product portfolio." Because it takes more than one capability to develop and deliver a product, a firm also has a number of in-house capabilities, which we describe as its "capability portfolio."[5] For example, Pfizer needed capabilities in medicine, medical testing, manufacturing engineering, and various other areas in order to deliver Exubera to its customers. The importance of the capability portfolio and how its constituent capabilities are recombined to produce products has been recognized for years.[6] However, what often goes unrecognized is that each of these capabilities in the portfolio is linked to each other in the network. Thus, a change in one capability will influence other capabilities and hence the effects of an innovation butterfly can be easily channeled from one part of an innovation firm, or even industry, to another.

---

[5] The importance of capabilities was first recognized by pioneering researchers studying the "resource-based view" of the firm, such as Rumelt (1984), Wernerfelt (1984), Dierckx and Cool (1989), Barney (1991), and Peteraf (1993). Lippman and Rumelt (2003) provide a good overview of this foundational work. Note that Prahalad and Hamel's contemporaneous (1990) concept of "core competency" overlaps significantly with their concept of a resource. Later, Leonard (1995) was among the first scholars to offer arguments of how to build capabilities internally within the firm, while Eisenhardt and Martin (2000) moved toward discussing how capabilities could be realigned over time by processes such as product development. Some researchers have built on these ideas to argue for a more dynamic view of capabilities. Others have extended this work by looking at some of the mechanisms that successfully realign capabilities at the business unit level (Martin and Eisenhardt 2010).

- Rumelt, D.P.: Towards a Strategic Theory of the firm. Alternative theories of the firmin. In: Robert Lamb (ed.) Strategic Management, pp. 556–570. Prentice-Hall, Englewood Cliffs, NJ, (1984).
- Wernerfelt, B.: A resource-based view of the firm. Strategic Manage. J. **5**, 171–180 (1984).
- Dierickx, I., Cool, K.: Asset stock accumulation and sustainability of competitive advantage. Manage. Sci. **35**(12), 1504–1511 (1989).
- Prahalad, C.K., Hamel, G.: The core competence of the corporation, Harv. Bus. Rev. **68**(3), 79–91 (1990).
- Barney, J.M.: Firm resources and sustained competitive advantage. JOM **1**, 99–120 (1991).
- Peteraf, M.: The cornerstones of competitive advantage: a resource-based view. Strategic Manage. J. **14**, 179–191 (1993).
- Eisenhardt, K.M., Martin, J.: Dynamic capabilities: what are they? Strategic Manage. J. **21**(10), 1105-1121 (2000).
- Lippman, S.A., Rumelt R.P.: The payments perspective: micro-foundations of resource analysis. Strategic Manage. J. **24**, 903–927 (2003).
- Sirmon, D.G., Hitt, M.A., Ireland, R.D.: Managing firm resources in dynamic environments to create value: looking inside the black box. Acad. Manage. Rev. **32**(1), 273–292 (2007).
- Martin, J.A., Eisenhardt, K.M.: Rewiring: cross-business-unit collaborations in multi-business organizations. Acad. Manage. J. **53**(2), 265–301 (2010).

[6] See the above note regarding Martin and Eisenhardt (2010).

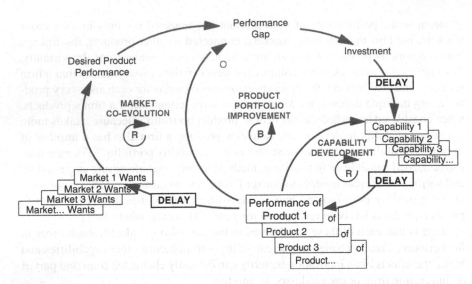

Fig. 3.1 Network of capability market system

In the automotive example, the improvements in electrical engineering capability to develop engine controls that could reduce emissions eventually strengthened the firm's ability to develop superior brakes through the development of antilock brakes, and ultimately this development also led to a weakening in the capability to engineer mechanical control systems. Hence, every one of a firm's capabilities is embedded in a network of relationships with all its other capabilities within its portfolio. For simplicity, we shall refer to these networks of capabilities, products, and markets and their linkage with each other as *structural complexity* to differentiate this from the temporal (or dynamic) complexity we described in the previous chapter. An illustration of such structural complexity through a network involving multiple products, markets, and capabilities is shown in Fig. 3.1. In this network, the total investment at a point in time may be fixed, and thus increasing an investment in Capability #1 (e.g., research of type 1 Diabetes) may come at the expense of Capability #2 (e.g., research in Type 2 Diabetes). A crucial point is that while human beings have innate intelligence and capabilities and, by extension, so do organizations, some capabilities and skills must be learned: they cannot be instantly acquired except under special circumstances. This has two implications. One is that a firm may have two or more similar capabilities that cannot be easily combined. For example, while approving the procurement of aircraft for Chinese airline, the local government may specify that a certain portion of the aircraft manufacturing tasks must be carried out in China. Or, the European Union may specify that a component must be procured locally for engine assembly. In such settings, the capability to manufacture the body of an aircraft in the USA may not translate directly into the capability to manufacture it in China or the European Union. In other cases, the capabilities may either be similar or complementary to one another, but their costs structure may differ. Hence, innovation organizations not only have to cater to different product markets, but they also may have to maintain development capabilities in different geographic locations.

We refer to such innovation systems as being *distributed* because many groups and individuals influence the firm's decisions. These influences have important implications for managing a firm focused on innovation.

The other implication revolves around the discussion of the automotive emissions revolution and the *principle of escalation of expectations* as discussed in the previous chapter; capabilities take time to develop—often much time. Because of this, it requires some amount of time before linkages between capabilities are understood and utilized by relevant development teams, and therefore their impact many not be visible to the customers immediately. To put it more succinctly, the distributed innovation is typically both dynamically and structurally complex.[7]

Unintended Consequences

Jay Forrester, Professor of Management at the Massachusetts Institute of Technology, has conducted numerous studies of complex systems such as urban land management and found that many of fixes for problems in such dynamically and structurally complex systems only serve to make the problems worse. He coined the phrase "policy resistance" to describe this phenomenon. An example of such resistance seen in his studies occurred in the Netherlands when government provided subsidies to the underemployed in a working-class neighborhood in The Hague to help improve the neighborhood's image and prospects. Counter-intuitively, this resulted in employed workers leaving the area, leaving the abandoned spaces to be filled by more underemployed people, worsening the area's image. This drove out more of the employed workers, who were better able to move because of their better finances, resulting in a vicious cycle. The planners had underestimated the effect of social mixture on migration and the good intentions of a city to relieve the pressure on underemployed ended with a neighborhood with an even poorer image and worse prospects.[8]

More interestingly, however, some fixes in complex systems actually work, but they have a nasty habit of creating problems in other parts of the system. That is, a combination of structural and temporal complexity leads to the possibility of

---

[7] For a discussion of distributed innovations, see:

- Anderson, E., Davis-Blake, A., Erzurumlu, S., Joglekar, N., Parker, G.: Managing Distributed Product Development across Organizational Boundaries, Chapter 10. In: Loch, C, Kavadias, S. (eds.) The Handbook of New Product Development Management, Butterworth–Heineman, Oxford (2008).

[8] Sanders, P., Sanders, F.: Spatial urban dynamics: a vision on the future of urban dynamics forrester revisited. Proceedings of International System Dynamics Conference (2004).

exchanges in sets of problems during an innovation process. For a concrete example of this exchange mechanism, consider river engineering. One of humankind's original "product engineering" achievements is the artificial taming of rivers that began with construction of dams along the Tigris and Euphrates of Mesopotamia from 5,000 to 6,000 years ago. River engineering served multiple needs or "markets," the foremost of which were to promote agriculture, prevent flooding, and later to facilitate trade. River engineering was extremely important for life and commerce—and still is—for a very large number of people and markets. Taming a river to provide adequate flood control is a highly complex process. One of the earliest and still heavily employed methods of flood control is levees (dikes), which are earthen embankments alongside a river that attempt to keep a swollen river from spilling over to the countryside or village. In fact, it could be argued that many capabilities of government actually arose as a function of river engineering. For example, part of the famous Hammurabi Code (c. 1650 B.C.), one of the earliest known codes of law, focused on flood control. In particular, it threatened to punish those who did not properly maintain the levees along a river that ran through their land[9]:

> If anyone be too lazy to keep his levee in proper condition and does not so keep it; if then the levee break[s] and all the fields be flooded, then shall he in whose levee the break occurred be sold for money [i.e. into slavery], and the money [from the sale] shall replace the grain which he has caused to be ruined.

While levees work well for small river swellings or for a short distance along a river, by stopping the spillover of water into upstream floodplains, they also increase the volume and speed of the water that passes downstream, hence increasing the potential for a flood downstream, and hence providing less warning of a flood than without a levee. As a countermeasure, during extreme floods, breaches in the levee system are often deliberately created to protect urban areas from flooding. For such a reason, a break was deliberately made upriver at Caernarvon, Louisiana, to protect the city of New Orleans during the Great Mississippi Flood of 1927. However, this solution created its own set of problems by ravaging the agricultural area known as St. Bernard's Parish near the levee break. The destruction caused an estimated 10,000 residents of St. Bernard's Parish to become homeless. Many were permanently displaced and eventually migrated to northern U.S. cities.

Sometimes the desire of one market or geographical site conflicts with another. For example, in efforts to make the Mississippi river safer and faster for the shipping market, the U.S. Army's Corp of Engineers has straightened the Mississippi's natural course since 1824. This also has provided the benefit of shortening the miles of levees that must be maintained.[10] However, these decisions have had many negative effects as well. Between the levees and the river's straightening, the volume of water that the Mississippi can absorb before flooding has been greatly reduced.

---

[9] Selections from the Code of Hammurabi (From King, L.W.: translation 1910; edition prepared by Gordon Patterson, PhD, Professor of Humanities, Florida Institute of Technology) www.wsu.edu/~dee/MESO/CODE.HTM. Retrieved 27 April 2010.

[10] Washington Post, Eilperin, J.: "Shrinking La. Coastline Contributes to Flooding," 8/30/2005.

Furthermore, millions of acres of wetlands have been destroyed due to straightening and dredging to allow larger ships to travel the river. In addition to ecological issues, removing the wetlands surrounding New Orleans to the south eliminated a barrier that protected New Orleans against hurricanes. Each acre of wetlands between New Orleans and the Gulf of Mexico absorbs some of the energy of hurricanes, slows down their winds, and reduces the storm surge that travels from the ocean into Lake Pontchartrain north of the city. Destruction of these wetlands contributed to the pressure to the levee system from Hurricane Katrina in 2005, which caused massive flooding of New Orleans and the loss of over 1,800 lives and devastated the economy and community life.[11]

Engineers have suggested diverting the Mississippi from its current course to create more wetlands (or at least to slow further erosion). However, this would likely result—at a minimum—in the displacement of people living, where the river would be diverted to and cause impediments to shipping. One of the key concerns of scientists in implementing this proposals is the possibility that they might make "irrevocable changes in the flow of the Mississippi River only to find out they caused more problems than they solve."[12]

These problems are clearly not unique to river engineering; they plague all complex systems. Returning to the innovation system, one of the greatest problems of managing innovation systems even in small firms is that managing numerous and varied customers, products and capabilities, and their associated links and their risks for disruption is a too complex task for the human mind, even when assisted by state-of-the-art computational capabilities.[13] Beginning with Adam Smith (1776) through Braha and Bar-Yam (2007),[14] decentralization has been identified a solution for managing risks in structurally and dynamically complex systems impacted by multiple streams of random events.

Given the potential for a rapid disruption anywhere within the markets that the firm's product portfolio is supposed to serve, innovation management theories also call for decentralized solutions. That is, planners assign most of the management responsibility to subunits, or departments, within firms that manage a small area of the business. Practically speaking, this means that most decision-making responsibility is given to innovation leaders who are responsible for producing at most a few, and sometimes only one, product. By eliminating the need for managers to confer with upper management on decisions and coordinate *operational decisions* with

---

[11] St. Petersburg Times, Waite, M., Pittman, C.: "Katrina offers lesson on wetlands protection," 9/5/2005.

[12] New York Times, Cornelia Dean, "Time to Move the Mississippi: Experts Say," 9/19/2006.

[13] See end-notes in Chapter 2 concerning this issue.

[14] Braha, D., Bar-Yam, Y.: The statistical mechanics of complex product development: empirical and analytical results. Manage. Sci. **53**(7), 1127–1145 (2007). Also see, Bar-Yam, Y.: Making Things Work: Solving Complex Problems in a Complex World. Knowledge Press, Cambridge, MA (2004).

other areas of the firm, decentralization indeed speeds up managers' reaction times to business disruptions in their area of responsibility. This is when the principle of exchange comes into play. By removing centralized planning, the need for explicit coordination between individual business units often increases. This has been shown by economics-related research any number of times.[15] The reason for this is that there is typically no "invisible hand" of market feedback capable of fulfilling this coordination role in a timely manner. Hence, there is the potential—and perhaps even inevitability—for two different business units to make decisions that, when combined, can be detrimental for the firm as a whole. For example, to save money and development time, many software firms develop software solutions that are customized to individual clients, but which are mutually incompatible. Such incompatibility, while attractive in the short term because it shortens development time, ends up being harmful in the long run. This is because when the business begins to grow, it will be unable to obtain economies of scale from the many incompatible solutions.[16]

The next chapter further examines the set of problems that arise from the decision to decentralize the management of the innovation system by introducing another source of innovation butterflies, that of human behavior.

---

[15] For instance, an analysis of policies for controlling the staffing and backlogs in a 2-stage service supply chain has been done. See:

• Anderson, E.G., Morrice, D.J., Lundeen G.: Stochastic optimal control for staffing and backlog policies in a two-stage customized service supply chain. Prod. Oper. Manage. 15(2), 262–278 (2006).

[16] Joglekar, N.R., Anderson E.G.: Global talent management: challenges of attrition, productivity and non-linear growth. In: Jain, K., Patil, A. (eds.) Proceedings of International Conference on Decision Sciences in Global Enterprise Management. McMillan Advanced Research Series, pp. 235–246 (2009).

Chapter 4
Providential Behavior

Take Care of Your Own Interests

*Controlling the specification of the product is one thing,
but the specification of the relationship with the supplier
is a whole other deal.*

Scott Palmer[1]

We have used the automotive electronics revolution example to describe the *principle of escalation of expectations*, whereby complex innovation systems act can amplify some innovation butterflies (seemingly small decisions or events) over time into tsunamis that drastically change the overall system in which the innovation firm operates. We have also described, using the examples from the pharmaceutical industry and river engineering, a *principle of exchange* in which, owing to the nature of problem solving in a dynamically and structurally complex system, any "solution" eventually creates its own set of problems. However, there is another aspect of the complexity that innovation teams must account for: the potential for misalignment between team's stated objectives and the choices made by individual team member or subteams. These misalignments are a final source for innovation butterflies as well as a potential drag on the efficiency of the innovation firm as a whole. We discuss these problems in this chapter in more detail using an example of a distributed project from a global software firm.

As discussed in the last chapter, individual contributors (such as a software architect) or subteams (such as a group charged with the development of a user interface in a video game development project) must have some freedom of action in order to be most effective in a complex environment. This results in the decentralization of the innovation firm's management authority to where it will best enable individual parts of the innovation organization to achieve their assigned goals. Different portions

---

[1]Quotation based on an interview conducted by the authors in 2006, while Scott worked in the purchasing/supply chain function at Sigma Tel, in Austin, TX.

E.G. Anderson and N.R. Joglekar, *The Innovation Butterfly*,
Understanding Complex Systems, DOI 10.1007/978-1-4614-3131-2_4,
© NECSI Cambridge/Massachusetts 2012

of the management structure will thus likely reside in different organizational and geographical locales with potentially different governance structures, decision rights, and review frequencies.

Innovation managers and workers are by their nature intelligent and creative and can be counted on to forecast the impact of their decisions upon the future. However, because they must focus only on part of the complex innovation system, they can only see part of the "big picture." Furthermore, each innovation worker or team has, in addition to whatever management direction they have received—their own goals, whether altruistic or opportunistic. This leads to the principle of providential behavior:

> Individuals (or groups) exhibit foresight and biases when managing their decisions related to complex innovation in distributed settings, which are based on their own perceptions about the future.

In other words, because of differences in goals, local perceptions, and individual biases, each individual innovation team—even completely altruistic ones—will rationally pursue its own agenda. The potential for misalignment between these agendas with each other and with that of the firm is not merely high, it is almost inevitable. The relevant question is not whether there will be misalignment of goals, decisions, and actions between the individuals within this system, but rather how great that misalignment will be. Recall the videogame example from Chap. 1. Once Nintendo has set the standards and created an accelerometer-based game, how would competition such as Sony and Xbox react? All their capabilities had been aligned for graphics, rather than motion-based competition. Aside from the overall wastefulness of misaligned efforts, even small misalignments create the potential for unanticipated "innovation butterflies," that can carry the entire innovation system into unexpected territory. The greater the misalignment is, the greater wasted effort and the greater potential for deviation of an innovation firm from its overall plan. Both of these effects interfere with the firm's ability to shape the innovation system over time.

Shortly, we discuss issues that increase these misalignments. However, there is a corollary to the *principle of providential behavior* that also must be considered. Because of the many balancing loops in the innovation system, over the long term the system will eventually correct itself by making the aggregate effects of these misalignments disappear one way or another. When misalignments create conflict between a firm's actions and market wants, the market wants will win out in the end because of the invisible hand embedded in the Market Co-Evolution Loop in Fig. 2.3. Either the firm will take actions to correct the misalignments, or eventually the firm will find itself losing market share. Importantly, however, the speed at which market correction occurs is for most innovation industries much slower than the speed at which the firm and its subunits can make decisions.

Hence, we offer a corollary to our observation about the principle of providential behavior: *Over the long run, the feedback loops of the innovation system seek to protect the interest of the customers by correcting severe misalignments of individual or group goals within the firm.* In this chapter, we describe the mechanisms that create the destabilizing effects and the corrective actions of the principle of providential behavior in more detail.

Free Will

Before attempting to avoid or wipe out destructive misalignments within the firm, we need to understand why these misalignments occur. To begin with, the people who work on innovation teams are special because they are charged with creating something that does not exist. Indeed, they may have been handpicked because they have some special skills (e.g., material engineering or creative marketing know-how) and they typically work in a group that has a culture and the commitment toward improving on the status quo. Many of these individuals have taken years of training in specialized areas such as engineering or market research. For them, working on an innovation project is not merely a routine job that requires executing well-defined tasks. While some tasks in an innovation project, such as project management effort, may be routine, other aspects of their task, such as brainstorming for solutions, is a creative process. The process of innovation has been characterized as a problem-solving contest, challenging and yet fun.[2] Even in "conventional" product development settings, such as in automotive development departments, this process involves a geographically dispersed team with members drawn from various departments of the firm and from its suppliers. This team tries to understand the customers' needs—both those they can articulate and those they cannot—and then provide a suitable solution. Each team member contributes his expertise toward formulating the problems surrounding the development of a new product and solving those problems. A typical shared goal for such a team would be improving the technical performance of a product feature, such as the efficiency of a windshield wiper blade or the amount of power that can be stored in a battery. The background of team members might differ depending upon the capability that needs to be delivered: mechanical engineers, tool and die makers, computer personnel, materials or chemical engineers, marketing managers, quality control, and parts suppliers. In addition to the technical skills, members bring in their imagination, foresight, leadership abilities, intelligence, charisma as well as their egos into the mix.

If the team meets or exceeds their performance goals, members may be rewarded with raises and promotions. If the product is a runaway hit, then developers can gain a great deal of respect from other innovation professionals along with raises and promotions. In extreme cases, when the product becomes well-known innovators may even gain public and professional recognition and, for a few, a rock star status. Steve Jobs at Apple, and in an earlier generation, Lee Iacocca of Ford and Chrysler became business rock stars by introducing a series of phenomenal product innovations. Some of these rock-star status individuals have been known for their passion, their egos, or often both while shaping the development of key products.

Since the development ecosystem is based on a profession that demands creativity, innovation professionals share some characteristics associated with people in other creative fields such as art and music. At a personal level, innovation

---

[2]Terwiesch, C., Ulrich, K.: Innovation Tournaments: Creating and Selecting Exceptional Opportunities. Harvard Business School Press (2009).

professionals oftentimes are driven by the chance to work on "cool" problems, and earn respect from their peers to more mundane rewards such as money, power, status or to lead a high quality life with the accouterments of wealth.[3] In other instances, they may be motivated by making a positive societal impact or, because of the team-driven nature of much innovation, preserving the jobs of their fellow professionals. More importantly, these goals differ from individual to individual. For example, almost all innovation professionals, whether engineers, CEOs, marketers, or programmers have begun at one time or another to identify with and nurture the products and services they are creating, almost as though these products are their own children. Such an emotional connection can shape their decision to act in an opportunistic manner (i.e., put their own goals ahead of the stated goals of their organization)—although it may not appear opportunistic to the innovation professional involved.

To complicate matters, in many development settings it is virtually impossible to track individual effort very closely either because the output depends on multiple people or the effort behind the output is intangible. For example, how does one track the number of unsuccessful attempts to solve a problem needed before one that works is found? Innovation professionals such as code writers in the software industry work on their own pet ideas or projects, sometimes at the expense of the overall departmental or organizational goal, because they are convinced that their pet project will bring in the right results for them *and* for their organization. In contrast, other innovators become invested in their pet projects primarily because they find the project inherently interesting or fun, because they have a hunch, or perhaps because they want to prove a point. In these cases, whether or not the pet project advances the firm's interests as a whole becomes at most a secondary concern. A few firms like 3M have recognized these tendencies on the part of innovation professionals and have attempted to channel this behavior by instituting a policy that allows up to 15% of the effort of individual developers toward learning, and or executing their pet projects.[4]

Alternative Perspectives and Biases

Because different innovation professionals draw upon different experiences, they necessarily have different perspectives resulting from the projects they have worked on and thus they may have different biases on the way the innovation tasks should

---

[3] See, for instance:

- Glen, P., Maister, D.H., Bennis, W.G.: Leading Geeks: How to Manage and Lead the People Who Deliver Technology. Jossey-Bass (2002).
- Stern, S.: Do scientists pay to be scientists? Manag. Sci. **50**(6), 835–853 (2004).

[4] For 3M's history on the 15% rule see: http://www.3m.com/us/office/postit/pastpresent/ history_cu.html.

be carried out. These perspectives or views will influence behavior, which we call exercising foresight and argue that such foresight is affected by personal biases, convictions, and experiences in much the same manner as the Jain parable about six blind men asked to examine an elephant to determine its nature by feeling different parts of its body. Each of them came to a different conclusion about the elephant's true nature.

The blind man who feels a leg says the elephant is like a pillar; the one who feels the tail says the elephant is like a rope; the one who feels the trunk says the elephant is like a tree branch; the one who feels the ear says the elephant is like a hand fan; the one who feels the belly says the elephant is like a wall; and the one who feels the tusk says the elephant is like a solid pipe.[5]

Such biases taken together with the attitude of the innovation professionals discussed earlier naturally produce different goals for each individual (e.g., a design engineer who is a parent of six children may be more interested in money and his family's quality of life vs. a fresh graduate who is interested in staying in a company for endless hours to work on "cool" projects). Thus, even when given an overall plan or "marching orders" to follow, the creative nature of innovation tasks allows individuals some degree of freedom in how to perform their tasks and thus some scope to pursue their own agendas based upon their individual goals and biases. Because of individual differences, the resulting individual agendas will not only misalign with each other, but also, in aggregate, with the innovation firm's plan as originally conceived. The only questions are: How great will the misalignment be? How will the ramifications of these different plans unfold over time across in a team, in a firm, and in the relevant industry? And how many butterflies will these differences generate?

The ongoing tsunami of distributed development, either through outsourcing (with team members drawn from multiple organizations), off-shoring (with team members from different countries and typically with different mother tongues and cultural lenses) or an open source arrangement has only served to increase the possibility of misalignment. Differences in individual biases can only be heightened when there are multiple organizational cultures involved, thus increasing the probability of misalignment. In addition, the time lags-associated communicating in distributed environments makes it more difficult to reconcile misalignments once they are detected.

Conventional wisdom suggests that one way to reconcile some of these misalignments in the innovation process is by inviting all the stakeholders to what goes by many names, but is often called a *charrette*—a collaborative session in which a group of developers attempt to come up with a shared vision and estimation for the

---

[5] Udana 68–69: We give a version of this well-known Indian tale from the Buddhist canon (http://www.co-intelligence.org/blindmenelephant.html), but some assert it is of Jain origin. It illustrates the Jain doctrine of Anekanta, the many-sidedness of things. http://www.cs.princeton.edu/~rywang/berkeley/258/parable.html.

tasks through prework and dialogue.[6] While there is no doubt a place for charettes in resolving misalignments, employing them without considering the dynamic and structural complexities of the innovation system is of little help.

To illustrate this idea, we next describe the impact of the principle of providential behavior on the estimation process for work in the software industry as observed by one of the authors.

The Estimation Dance

> The Admiralty asked for six battleships, the Cabinet proposed four, and we compromised on eight.
>
>   Winston Churchill [1914]

SofTex, a global software development firm with nearly 5,000 employees has deep experience in developing e-commerce solutions that integrate legacy (i.e., currently existing) information systems and processes across multiple companies into a unified digital marketplace.[7] One of its innovation projects involved integrating a customer-facing website with an existing (or, in software jargon, "legacy") software application. Two teams were involved. One team was responsible for developing the web front-end and was composed of 20 in-house SofTex employees. Another group of nearly 60 developers, responsible for modifying the legacy software to integrate it with the web front-end, joined the project through an outsourcing arrangement with an outside, overseas supplier firm with a high degree of experience. Some of the supplier's personnel worked on site; others performed their work while remaining overseas. Fortunately, both SofTex and its supplier were quite sophisticated in project management and coordination.[8] Prior to execution of this particular SofTex project, the team management (a group of eight individual from the principle firm and the supplier) first estimated its scope by first identifying development and integration tasks and then splitting them among multiple sub-teams, some located in the USA, and the rest at an offshore development site. A typical SofTex product contains about 10,000 function points, a standardized measure of functionality that an information system provides the customer. While function points are standardized with respect to functionality, they can vary greatly

---

[6] The term "charrette" is commonly used in the urban planning and in the construction industry. See http://www.charretteinstitute.org/charrette.html.

[7] This case is based on work carried out at SofTex by one of the authors. The name of the firm and the details of the project have been altered to preserve anonymity. Allied of the technical details are available in:

- Joglekar, N.R., Yassine, A., Eppinger, S.D., Whitney, D.E.: Performance of coupled product development activities with a deadline. Manag. Sci. **47**(12), 1605–1620 (2001).

[8] The supplier's development process was certified to comply with a "CMM level V" rating, the highest possible based on the industry-standard System Engineering-Capability Maturity Model (http://www.sse-cmm.org).

with respect to the difficulty of implementation. At SofTex, most function points varied between a moderate and extremely difficult degree of implementation. SofTex (and the supplier) team began to model the targeted product performance initially using a Unified Modeling Language (UML), a simplified lingua franca used by programmers to detail their work.

The leadership team knew that coordinating two teams of innovation professionals working on a project in parallel would be a time-consuming and difficult process. A SofTex project, which integrated a web-based front-end onto a legacy application, illustrated some of the reasons for these problems. For example, it was anticipated that any work done by one group in parallel with a second *off-site* group was bound to cause the second group to have to redo some of its previously completed work because of bugs at the interface of the two groups' work. This is known as *rework* and is unfortunately an all too common phenomenon in the software industry.[9] However, in the SofTex planning effort, both teams made very poor estimates of the impact of their work upon the other's. In particular, code developers in both teams systematically underestimated the amount of rework they would create for the other teams, while overestimating the impact of the other team's actions in creating rework for themselves. This could be attributed to prior experiences as well the their desire to create a buffer to guard against the uncertainty around the work that they would be responsible for in the future, based on the initial tasks carried out by someone else.

The supplier team also gave an estimate for the resources needed for their own work that was significantly lower than the system architect's estimate. SofTex managers felt that this gap was due to the supplier firm's lack of experience on this particular technology and their desire to gain this experience, even if it meant that they had underbid on this particular aspect of the task. In other words, the goals for the supplier team differed from those that would most benefit the project as a whole. Equally interesting, the estimate given by the in-house team also differed from the system architect's, but by being significantly higher rather than significantly lower! The mere fact that there was any disagreement in the various estimates, given SofTex's sophisticated and data-rich environment, underscores the importance of biases in driving the *assumptions* behind a specific team's estimates. A rationale for the estimation gap was suggested by SofTex's management in this instance: the in-house team simply padded their numbers to hedge against uncertainty. In their experience, this was a natural—though suboptimal—consequence of the fact that in-house teams were rewarded for completing the project as closely as possible to the estimated time, rather than completing the project as quickly as possible.

Playing games with the estimation process is, therefore, a common malady for almost all projects of any type and complexity. One of the authors is familiar with management of projects of enormous technological complexity at one Fortune 50

---

[9] For a comprehensive discussion of this type of rework in the software development setting, see:

- Abdel-Hamid, T., Madnick, S.: Software Project Dynamics: An Integrated Approach. Prentice Hall, Upper Saddle River, NJ (1991).

firm from the energy sector, which we shall refer to as ProjCo to mask its identity. ProjCo has grappled with the uncertainties of completion time estimate by asking for probabilistic estimates for project completion that are expressed calculating 5th, 50th, and 95th percentiles of completion times at the beginning of each project. This initial 50th percentile estimate is often referred to as a project's 50/50 estimate because about 50% of projects should be completed before their 50/50 estimate and 50% should exceed this estimate. However, more than one executive at this firm observed that about 70–80% of their firm's projects are completed before the 50/50 estimate. In other words, the 50/50 estimates were "padded" sufficiently so that about 20–30% of the firm's projects were beating their 50/50 estimates that should not have. Most executives at the firm believed that this "padding" was due to the fact that management punished project managers who completed their projects "late" relative to the initial 50/50 estimate extremely severely when they reviewed overall performance and awarded bonuses. In essence, the 50/50 estimate had become the target by which project manager's performance was judged.

Padding targets is not unique to SoftTex or ProjCo. Long and intense negotiations have been documented in detailed studies of engineering projects by Buchiarelli.[10] In our view, given the structural complexity of innovation projects and the inherent uncertainty around the project scope, smart managers tend to hedge against the probability of emergent uncertainties (i.e., innovation butterflies) in order to protect their own and their teams' reputations and careers. Most project managers and innovation professionals know that, at the end of a project, their performance will be judged against the expectations set up at the very beginning. Is there a better place to start this type of hedging than during the very process that sets expectations?

Foresight and Negotiation

Next, we examine the impact of such distortion on the market-capability dynamics. Managers may start the target-setting process by "padding" into their estimates up front and asking for additional resources, when estimation uncertainty is large.[11] In effect, this behavior distorts the planning process for each generation of project captured in Fig. 4.1. Since each subteam team is also likely to play by similar rules—each individual does his or her own estimate padding. This leads to negotiations before the effects of organizational learning shown on the right-hand side of Fig. 4.1 can be set into motion. Figure 4.2 shows the typical evolution of performance targets based on a set of conversations about the evolution of agreed upon performance

---

[10] Buchiarelli, L.L.: Designing Engineers. MIT Press, Cambridge, MA (1994).

[11] For a review on the literature on uncertainty estimation and the selection of desired performance levels during software planning, see Boehm, Abts and Chulani (2000). We note this literature recognizes that complexity matters, however, most software development practices, except agile development (which we cover in Chap. 5) are not geared to handle emergent complexity.

- Boehm, B., Abts, C., Chulani, S.: Software development cost estimation approaches—a survey. Ann. Softw. Eng. (2000).

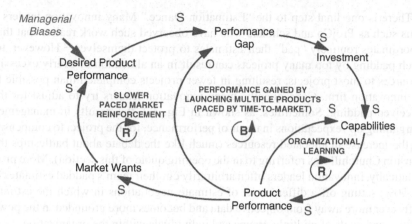

Fig. 4.1 Setting expectations before organizational learning takes effect

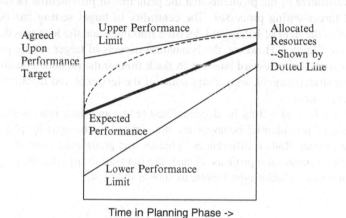

Fig. 4.2 Evolution of desired performance during planning phase

targets during the negotiations between management and the team(s) responsible for delivery at the onset of projects in at ProjCo. It is worth noting that the setting of performance target is not a onetime event. This dance repeats itself in each subsequent generation, because the market place ups the performance ante.

Thus, the "estimation dance" is a risk assessment process in which each participant subteam comes to the negotiation table and asks for resources, while reporting on a wide list of uncertainties at the outset. Negotiations during the planning phase allow this team and the planners to consider a heterogeneous set of issues and possible events and to assess their probability and risk-impact levels. The participating teams fully understand that their performance in the future will be measured with these targets as the benchmark. That is, the perception of the team's success within the parent organization, and their self image, are based on how well they are doing against the bench mark set by the estimation process. Hence, individuals and groups *exercise foresight* in setting desirable performance targets.

There is one final step to the "Estimation Dance." Many innovation leaders at firms such as ProjCo and scholars who have observed such work realize that their subordinates routinely "pad" their estimates to protect themselves.[12] However, too much padding by too many projects can result in an allocation of overly excessive resources to most projects, resulting in fewer projects completed than possible by the innovation firm. Consequently, many innovation leaders try to adjust for this perceived padding. Sometimes, as shown in Fig. 4.2, this results in management "bargaining up" expectations in terms of performance, for the project to compensate for the increase in allocated resources (much like the debate about battleships that Winston Churchill was referring to in the opening quote of this section). More problematically, innovation leaders often arbitrarily cut their team's padded estimates by 20–50%, setting off a different set of estimate negotiations in which the estimate drifts ever more away from objective data and becomes more grounded in the power politics and goals of multiple groups and individuals within the organization.

Not only are the distortions of the estimation dance important in themselves, but they are illustrative of the problems that the principle of providential behavior creates in all target-setting processes. The centrality of target setting has been well understood in the field of behavioral decision making since the idea was developed by Cyert and March.[13] However, the behavioral aspects of target setting processes, such as the team's bias toward building in slack time or the planner's desire to seek and enforce stretch targets, are perhaps some of the less explored frontiers of innovation management.

Distortions in goal setting by decentralized teams and their management based on principle of providential behavior are not limited to the start-up phases of an innovation project. Rather, differences in biases and goals exist over all aspects of the lifecycle of innovation projects. Hence, the innovation firm and its projects can be caricatured by a Calder-type mobile as shown in Fig. 4.3.[14]

---

[12] Ford, D., Sterman, J.: The Liar's club: concealing rework in concurrent development. Concurr. Eng. Res. Appl. 11(3):211–219 (2003).

[13] Cyert, R.M., March, J.G.: A Behavioral Theory of the Firm. Prentice-Hall, Englewood Cliffs, NJ (1963).

[14] It is common practice in the literature on product architecture to represent interconnections using a hierarchical structure (e.g., Alexander 1961; Simon 1969; Clark 1986). For a companion discussion of the technical problems of hierarchical product and portfolio planning and the underlying uncertainties, see Anderson and Joglekar 2005).

• Alexander, C.: Notes on the Synthesis of Form. Harvard University Press, Cambridge, MA (1964).
• Clark, K. B.: The interaction of design hierarchies and market concepts in technological evolution. Res. Pol. 14(5), 235–251 (1985).
• Simon, H.: The Science of the Artificial. MIT Press, Cambridge MA (1969).
• Anderson, E.G., Joglekar, N.R.: A hierarchical product development planning framework. Prod. Oper. Manag. 14(3), 344–361 (2005).

For a discussion of how to create a Calder mobile, see http://www.instructables.com/id/How-to-Create-a-Calder-esque-Mobile/.

Fig. 4.3 Balancing goals across Project's Calder mobile

This particular mobile involves certain project subteams who are sitting in rela tively stable spots, and others who know that their position is precariously balanced, and at least two managers who are operating with a blindfold because they simply cannot estimate the amount of uncertainty inherent in their tasks either because of their novelty or because their work depends on what others might do. Anyone who has played with this type of mobile (like those that sit above many baby's cribs) knows that this complex system is precariously balanced but balanced nonetheless. While they are generally stable if left alone, it is very easy for a part to tip if one tries to adjust the mobile by hand (or if even a gust of wind hits it from the motion of a door, that is being closed, or a window being opened). This instability caused from random or not so random motion represents the butterfly effect in motion.

Scope Creep and Other Pathologies

Other behavioral factors can also cause the innovation butterfly to flutter its wings. Even after the planning stage, so long as the project team continues to innovate and make late changes in specifications, many events will occur that can create butterflies. Many project managers have encountered projects in which the scope of the project increased over the course of its development.[15] Unfortunately, this all too often

---

[15] For a review on the literature on planned vs. unplanned rework in a single project, see Safoutin and Smith (1998).

- Safoutin, M.J., Smith, R.P.: Classification of iteration in engineering design processes. Proceedings of the ASME Design Engineering Technical Conference: DETC98/DTM-56723 (1998).

results in litigation to decide what party is responsible *for paying for project cost overruns or delays*.[16]

As we have learned from the Wii case study in the introduction, the uncertainty associated with the nonroutine aspects of innovation tasks, and the occasional arrival of the innovation butterfly flapping its wings, causes individuals on a project to often deviate from previously agreed-upon goals. To further confuse matters, because of the complexity of the innovation system, it is difficult to ever pinpoint the status of a project, even without the introduction of scope creep. These two factors taken together make the decentralized management innovation projects and portfolios even more susceptible to opportunistic behavior by individual innovation professionals or subteams. One example, described by Ford and Sterman, is a phenomenon that they label "liar's club" in which project reviews lead to a standoff because managers hide their true progress status (particularly by avoiding being the first to report a "bug") because they know that other managers will hide their own status as well.[12] Eventually all the hidden bugs show up, but owing to the complexity of the innovation system, it is difficult to pin down the cause of the problem (or the butterfly event, in our terminology) to any one individual or subteam. Examples of such opportunistic behavior are more prevalent in distributed settings where the economic incentives for various collaborating teams are misaligned. For instance, self-serving tactics to extend contracts by professional services firms have been documented in great detail.[17]

In any one project, given all the complexity and distortions in measuring progress, only a few recourses are available for managers to automatically discipline people who are "behaving badly," i.e., exhibiting the human foibles of opportunism, bias, deceit, and other suboptimal behaviors. The ability of innovators to exercise free will not only creates butterflies but also amplifies their impact in destabilizing the intended path of progress within an innovation system.

Multiple Projects: Determinism

On the face of it, the presence of multiple projects only makes the management of the innovation more challenging. As an example of this, some researchers have focused on deadline effects and the decisions of individual innovation professionals

---

[16] Cooper, K.G.: The rework cycle: vital insights into managing projects. IEEE Eng. Manag. Rev. Fall Issue, 4–12 (1993).For a discussion of unplanned rework during Boston's "Big Dig," see:

- Lewis, R., Murphy, S.: Artery errors cost more than $1b. http://www.boston.com/news/specials/bechtel/part_1/. (2003).
- Ross, C.: Dig firms knowingly hid million$ in gaffes, http://news.bostonherald.com/localRegional/view.bg?articleid=65138. (2005).

[17] Rockart, S.: How do professional services firms compete? Proceedings of 19th International System Dynamics Conference, Atlanta, GA (2001).

to "play hero" by firefighting in order to deliver unrealistic deadlines for a current innovation project, at the expense of the progress of other later-deadline projects.[18] Portfolio managers play along by shifting resources toward the "hot" projects, and yet they are not able or willing to delay or cancel other projects. The net result, as observed in many innovation settings, is that innovation firms often find themselves stuck in a never-ending cycle of firefighting and under-performance. This might also be yet another reason to have the foresight to seek for additional resources during the estimation dance described above.

However, the importance of market mechanisms cannot be neglected, i.e., how well a product is received, since often across successive development cycles, products are introduced to the marketplace and market feedback is observable in terms of sales figures. This feedback can become an integral part of the decision loop, such that the development team's free will is tempered by the built-in *determinism* brought about through consumer action within the system. As an example, recall our earlier example from the automotive setting in Chap. 2 when in succession: mechanical braking systems were deemed insufficiently safe; they were then replaced by electronically controlled antilock braking systems (ABS), made possible by the electronics engineering capability originally developed to reduce emissions; the market eventually accepted ABS; finally, ABS systems were expanded into traction-control systems, which allowed drivers to not only brake, but also to actually drive and even accelerate on slick surfaces.

Given the importance of planning over multiple product cycles, how do the motivational aspects of the innovation structure discussed in the previous section play out over multiple projects? The interval between successive product launches, also known as the *time to market*, can be as short as a few weeks to a few months for software releases or hardware upgrades in the information technology industry. In other industries, such as toys and clothing, the launch schedules work on a yearly cycle, with developers aiming to position their products for peak demands during the holiday season at the end of the year. Moreover, whatever the industry, the market feedback lags behind the launch of the product by several months or even years. Hence, at the end of each development project, it is difficult to tie economic incentives such as bonuses for teams or individuals to the marketplace success or failure of a project. In many instances, these bonuses are instead tied to process measures, such as the time it takes to complete the project and release it to the market, instead of market success such as observed sales.

Across multiple development cycles, typically over 12–36 months in many industries, innovation leaders are in a better position than their subordinates to observe market response and adjust their investments in capabilities that they perceive to be critical. So a firm might keep or build up a particular design style in toys or develop or enhance a software graphical user interface (GUI) because managers believe that such a capability is valued in the marketplace. Similarly, innovation

---

[18] Repenning, N., Gonclaves, P., Black, L.: Past the tipping point: the persistence of firefighting in new product development. Calif. Manag. Rev. (2001).

leaders may divest and reposition skill sets when some capabilities are deemed to be undervalued or obsolete in the marketplace. In other instances, such divesture occurs because much cheaper alternatives are available elsewhere, typically overseas.

Innovation professionals do not have the same level of access as their managers to marketplace feedback and in many instances they are not privy to the detailed factors behind managerial decisions. In the short term, they may receive economic incentives such as bonuses or may see others rewarded instead. In the long run, they observe that some capabilities are being built, while others are ignored or become stagnant. For instance, in the SofTex case mentioned earlier, software engineers may have decided that the use of database query languages and data mining techniques for the legacy back-end software was passé because the firm invested much more in modular web-based software solutions, such as IBM's Websphere, to support the front-end. Innovation professionals understandably tend to see these decisions as reflections of upper-management's intent.

Innovation professionals assess trends across successive projects based not only on short-term economic considerations such as bonuses, but also through their relationships with the teams and suppliers, just as the quote by Scott Palmer at the start of this chapter indicates. Other motivations are often linked with potential opportunities to do "cool" projects, keep their careers afloat, visit interesting places, or learn new tricks within their domain of expertise, plus their desire to belong to a "winners" circle' in terms of relevant capabilities.[19] Additionally, most innovation professionals and teams are looking for a sense of belonging and a place of pride within their community of technical peers. This is as true for software engineering teams working in Silicon Valley as it is for automotive development teams working for Honda in the USA or for their counterparts working for Honda in Asia. Perhaps, the desire for recognition and pride plays an even larger role in their willingness to exercise free will when these engineers live in closely knit residential communities or in a company town surrounded by peers and their families. The buzz about creative contributions and positive impacts on performance by individuals and their teams are evident beyond the workplace and are talked about by friends and families alike. Again, such peer respect and community appreciation is as evident in a residential subdivision full of Intel employees in Santa Clara, California, as it is in high-rise buildings in Changwan—home for some of the cutting edge consumer electronics production in Korea—or in gated communities full of software developers in Bangalore, India. The bias for exercising free will in promoting decisions toward building capabilities and making it into a winners' circle, is a virtuous turn of the performance-capability cycle that builds on itself.

In special settings such as the Silicon Valley, engineers may be willing to take on further risks and forego their secure jobs and bonuses due to the lure of stock options and the chance to create something dramatically different and innovative.

---

[19] Siemsen, E.: The hidden perils of career concerns in R&D organizations. Manag. Sci. **54**(5): 863–877 (2008).

However, their professional relationships, social networks, and peer appreciation remain in place. These networks admire risk takers. There is ample evidence to show that startup engineers and managers also form teams and communities that contribute to a serial set of startup successes.[20] Even in startup settings, the urge to exercise free will in the creative process goes well beyond economic motivations such as stock options. Entrepreneurs wish to build positive or virtuous feedback loops that are observed by peers, and by venture capitalists, across a series of development projects.

So far, we have argued that beyond the obvious short-term economic incentives, the innovators' work in the trenches is influenced by their desire to capture the positive effects that are generated by long-term build up of capabilities, through virtuous cycles of successes. But no cycle can last forever and there must be limits to market growth; even dominant architectures are susceptible to attacks from a disruptive technology,[21] and a virtuous cycle can turn vicious once diminishing returns on capabilities set in or the signs of customer divesture into a particular technology become evident. Seasoned innovation professionals fear this loss of relevance as much, if not more, than the loss of a year's bonus.

Multiproject Pathologies

Another set of unintended consequences of the evolution of a project or a series of projects results in two sets of issues: turf building and the emergence of disruptive technologies in the marketplace.

A major source of inefficiency and risk based on the evolution of capability is the tendency of individuals to build "turf," i.e., their own little zones of comfort and influence. A typical side effect of turf building is the likelihood that developers either get complacent or focus on defending their turf; in either case, they lose sight of the market. This often manifests in the curse of "lookalike" products in which two different project teams each change the functionality of their projects to appeal to the other's designated market. Thus, the firm ends up with two products, both of which appeal to both markets. This wastes at least one project's worth of resources, while potentially leaving a gap somewhere else in the portfolio. Instances of infighting between divisions of the big three Automakers provide legendary examples of this phenomenon in the auto industry,[22] and has resulted in much of the product duplication that has plagued that firm. Such duplication in development and distribution effort is costly, and often leaves the end customers confused about how

---

[20] Zahra, S.A., Jennings, D.F., Kuratko, D.F.: The antecedents and consequences of firm-level entrepreneurship: the state of the field. Entrepreneurship Theor. Pract. 24(2), 45–66 (1999).

[21] Christensen, C.: The Innovator's Dilemma: The Revolutionary Book that will Change the Way you do Business. HarperCollins, New York (1997).

[22] Wright, J.P.: On a Clear Day You Can See General Motors: John Z. DeLorean's Look Inside the Automotive Giant. Wright Enterprises (1979).

to select an offering from product portfolio. Another example comes from our fieldwork in which two marketing managers, one from a computer networking division and another from a software product division, each tried to pitch their own individual solutions to the same customer. These product duplications are not an isolated incident of self-serving behavior. In the aggregate, such self-serving behavior cuts much deeper than an occasional launch of similar products resulting in a waste of resources. For example, another aspect of turf building around an individual team is to monopolize control over a particular innovation capability, when it would most efficiently be scattered in multiple places across the innovation firm.[23]

A related major source of risk in developing a capability-reinforcement-based strategy is the potential for being blind toward emergent disruptive technologies. For example, Christensen[24] describes the mechanisms that lead to the failure on part of a succession of relatively expensive storage technologies, whose sales were disrupted by the acceptance of low-end substitutes by the consumers. Christensen's careful description such disruptions in the data storage industry usually started as butterflies, small changes in unrelated markets, that gathered steam to create disruptive tsunamis.

Over time, these and other manifestations of self-interested behavior by decentralized decision makers are eventually penalized by disenchanted consumers. That the decline of General Motors since 1970 is a function, in part, of the misaligned self-interested behavior of decentralized decision makers is well known and heavily documented. While GM invented some ground breaking technologies such as OnStar and housed a highly trained and motivated work force over the past 30 years, it has also suffered from numerous instances of the curse of self interested behavior including lookalike products, turf-building, and estimation dances, that finally resulted in its nationalization during the economic down turn of 2009–2010.

The three principles of escalation of expectations, exchange, and providential behavior, when taken together, put innovation leaders into a dilemma when they attempt to shape the innovation system. By decentralizing control, they permit their employees to more efficiently cope with the complexity of the innovation system resulting from the principle of escalation of expectations. On the other hand, this very decentralization activates the principle of exchange, by permitting the opportunistic and other dysfunctional behavior described in this chapter. Given decentralization, the principle of providential behavior guarantees that such dysfunctional behavior will occur and, furthermore, that it will eventually be punished by the market. How can innovation leaders reap the rewards of decentralization while minimizing the effect of opportunistic behavior? In the next three chapters of the book, we discuss a host of alternatives strategies for how leaders can offer a beneficent guiding hand while coping with the complexity of the innovation system.

---

[23] For vivid descriptions of such turf building and infighting in the automotive sector see Halberstam (1986).

• Halberstam, D.: The Reckoning. Avon Books (1986).

[24] Christensen, C.: The Innovator's Dilemma: The Revolutionary Book that will Change the Way you do Business. HarperCollins, New York (1997).

Part II
Agile Product and Portfolio Planning

> *Before I sink into the big sleep, I want to hear ... the scream of the butterfly.*
>
> Jim Morrison

So, how do we manage the challenges of innovation systems arising from the principles of *Escalation of Expectations, Exchange, and Providential Behavior*? Recall that the principle of the *escalation of expectations* occurs because an innovation firm develops its portfolio of capabilities over time based on the market's past response to its products, while simultaneously the market changes its expectations for future products based upon the products offered by the firm and its competitors. This explains why innovation systems are almost lifelike in their behavior, occasionally magnifying seemingly small innovation butterflies into tsunamis of industry-wide change. The principle of *exchange* describes how the complexity and connectedness of the innovation system result in the emergence of unanticipated outcomes from any managerial actions. Finally, the principle of *providential behavior* describes how individuals and teams in decentralized management settings will invariably produce goal misalignment, which is both a drag on efficiency and yet another source of innovation butterflies. The three principles, taken together, create the potential for innovation butterflies that result in difficult-to-manage emergent phenomena.

How do organizations adapt to the trinity of challenges just described so as to effectively manage innovation butterflies? Many approaches and experiments have been taken in practice. Much of the received wisdom regarding these experiments has been written about, particularly with respect to managing individual projects. For instance, the "agile" software development methodology (which we discuss in the next chapter) teaches project managers to make rapid adjustments in order to manage the principles of escalation of expectations and exchange so as to reduce the likelihood of "runaway projects" that are severely over-time, over-budget, and—in some cases—unlikely ever to be completed. The essence of the agile method is to hold rapid reviews to ensure individual team-member alignment and halt tasks that might result in runaway projects—in which the rate of discovery of new rework tasks exceeds the ability of the project team to resolve them—or other undesirable emergent phenomena resulting from innovation butterflies. On the flip side, agile project management can more effectively exploit desirable innovation butterflies to create emergent phenomena that benefit the firm than can traditional methods of project management.

In contrast, the received wisdom for managing a *portfolio* of innovation projects by leveraging a set of capabilities to cope with innovation butterflies, however, has many gaps. This is unfortunate, because the portfolio level is a point of high leverage within the innovation system. In this section, we introduce a set of concepts to address these gaps by illustrating how innovation leaders can achieve agility and flexibility across a portfolio of evolving projects.

As has been discussed earlier, the three principles described in the first section of the book turn the management of innovation into a dynamically complex process that can transform seemingly minor innovation butterflies into tsunamis of change. At an early stage, before these butterflies begin to grow their wings, an innovation team typically finds it difficult to judge if the butterfly effect will become an opportunity or not. Often innovation butterflies can show up in swarms, clouding the innovation manager's vision. Combined with all the other uncertainties, ambiguities, and individual biases in the innovation system, they can become an almost impenetrable "fog." While such a system sounds impossible to manage, it may be possible that an innovation team, if it moves quickly enough or is skilled or lucky enough, can actually drive the innovation process in a direction that favors their firm and its capabilities, much like Apple and Nintendo. In this section, we describe how a firm can break through the "fog" inherent in innovation systems to do this by drawing upon ideas of *pattern recognition* and *scaling* from complexity science and the ideas of *pattern recognition* and *maneuver warfare* from military science.

One central lesson in complexity science is that the effects of complexity can be identified and managed by looking for patterns of behavior at a higher level of aggregation. For example, tracking the behavior of a gas (like a cloud of steam) in terms of its temperature, pressure, and volume, is a hopeless task if one tries to track the movements and properties of individual molecules, each of which is essentially random. However, if one can "scale up" and track the molecules as an aggregation instead, then the properties of the aggregation can be described by well-known, very simple physical laws that govern all gasses. Thus, by ignoring the behavior of *individual* molecules, tracking the behavior of aggregation as a whole is highly simplified. We argue that innovation leaders can create a similar simplification and achieve greater leverage by considering their innovation projects and capabilities as portfolios rather than trying to manage the details of each individual innovation project. Similar scaling ideas apply in warfare when one moves from command of a squad of soldiers to that of a brigade, division, or army.

We will continue to push on this military analogy because, while warfare is not a perfect analog of the innovation system, it does offer a number of instructive parallels for the management of innovation systems. Skilled military leaders and organizations have, under certain circumstances, navigated some of the most chaotic (both mathematically and otherwise) systems known to humankind. Carl von Clausewitz, a military theorist whose ideas have heavily influenced the organization of many modern military systems, summed it up nicely with two concepts.[1] One was "fog" (or sometimes "fog of war") which he described as the prevalence of myriad random and unforeseeable events that obstruct any semblance of linear progress in a landscape clouded by uncertainty. In Clausewitz's words, "Many intelligence reports in warfare are contradictory; even more are false, and most are uncertain." Another is his concept that "Everything in *war is very simple. But the simplest thing is very difficult* [author's emphasis]. These difficulties accumulate and create a kind of friction." The result of this "friction" is to make orchestrated maneuvers extremely difficult. Clausewitz used the following analogy to describe the effects of friction, "Just as a man in water is unable to perform with ease the most natural and simplest movement, that of walking, so in War, one can attain even mediocre results only through extraordinary effort." Friction includes the impact of both unexpected random events and the dynamic and temporal complexity embedded in the system that is warfare. In short, "friction" is the process that makes the execution of detailed, intricate plans essentially impossible and "fog" refers to ambiguity under which decisions must be made.[2] The parallels between managing the complex systems of warfare and that of innovation are striking. As any project manager knows, there exist direct analogs of both "friction" and "fog" in the innovation system, which are created by the three principles of *escalation of expectations, exchange,* and *providential behavior*—as well as a number of other factors. Not the least of these is the presence of active competitors in the innovation system, who—while not precisely enemies in the military sense—will, in their own quest for market success, almost inevitably disrupt the innovation firm's plans for profit and growth.

How have military organizations dealt with competitive threats and achieved success? The classic answer is through using the element of surprise at an opportune place and time when the enemy is weak. However, the gap between recognizing such an opportunity and acting upon it in a system obscured by "fog" and impeded by "friction" is huge. Historically, the most effective armies have generally (1) trained to create a high level of *preparedness* and (2) empowered their subordinates to take *independent* but *coherent* actions that can contribute to shape the course

---

[1]Clausewitz, C.v., Howard, M., Paret, P. (eds.): On War [Vom Krieg] (Indexed ed.). Princeton University Press, New Jersey (1984) [1832].

[2]It is well documented that certain organizations (as pointed out by Clemons and Sanatamaria 2002, to whom we are deeply indebted) have survived and even thrived in the chaotic world of warfare. An interesting example of this philosophy is the famous 90° turn by the Third U.S. Army under General George S. Patton Jr. in World War II's Battle of the Bulge. Other examples include Napoleon's lure of Austrian and Russian armies into a mistaken attack on his right flank (allowing the French to crush the weak Allied center) at Austerlitz, and General William Sherman's unexpected move to cut his army off from its own supply line during its March to the Sea in the American Civil War.

Clemons, E., Santamaria, J.A.: Maneuver warfare and competitive strategy in rapidly evolving markets. Harv. Bus. Rev. (2002).

of battle in their favor. These ideas of preparedness and independent, but coherent actions are the core principles underpinning *maneuver warfare*. We discuss tools that, taken together, can create an analog of maneuver warfare, which we term *maneuver-driven competition* that is suited for the endeavor of innovation management.[3]

This section is organized as follows. Chapter 5 describes agile project and portfolio management and introduces the core concepts behind *maneuver-driven competition*. Chapter 6 outlines how to support maneuver-driven competition through strategies for managing innovation portfolio risk, and Chap. 7 focuses on supporting maneuver-driven competition through shared resources and processes.

---

[3]Maneuver warfare has been viewed as the basis of success of modern military organizations including the U.S. Marine Corps (USMC) and the Israeli Defense Forces. The concepts of maneuver warfare are rooted in the theories of Sun Tzu, and the French and German military reformers of the Revolutionary and Napoleonic wars, Carl von Clausewitz, Helmuth von Moltke, J.F.C. Fuller, Mikhail Tukhachevsky, Gunther Blumentritt, Martin van Creveld, and reached fruition with legendary briefings of US Air Force Colonel John Boyd.

We note that the ideas of adaptation, agility and maneuvering have been explored in the management strategy literature. In most instances, this literature addresses aggregate firm level strategy, rather than project or portfolio level innovation strategy. See for instance:

Bettis, R.A., Hitt, M.A.: The New Competitive Landscape. Strat. Manag. J. **16**(Summer) 7–19 (1995).

Ilinitch, Y., D'Aveni, R.A., Lewin, A.Y.: New organizational forms and strategies for managing in hypercompetitive environments. Org. Sci. (1996)

The idea of agility, particularly at the portfolio level can also be mapped into the concept of "Clockspeed," developed by Professor Charles Fine:

Fine, C. H.: Clockspeed: Winning Industry Control in the Age of Temporary Advantage. Perseus Books, Reading, MA (1998).

Chapter 5
Agility and Maneuver-Driven Competition

Recognize Patterns and Adapt Projects and Portfolios

> Each element can move, shift or sway back & forth in a
> changing relation to each of the other elements in the universe.
> Thus, they reveal not only isolated moments, but a physical law
> or variation among the elements of life. Not extractions, but
> abstractions. Abstractions which resemble no living things
> except by their manner of reacting.
>
> Alexander Calder[1]

Martin van Creveld in *Command in War* (*1985*) points out that, to cope with the "fog" and "friction" inherent in warfare, ideally a military genius would be in charge of every individual subformation at every critical point during a battle or campaign.[2] (A subformation refers to atomic units within larger military group, e.g., 8–10 solders form a squad, 3–5 squads form a platoon, many platoons form a company, many companies constitute a battalion, and so on.) Obviously, van Creveld continues, this is impractical as geniuses are in short supply and critical points are difficult to identify in advance, even for geniuses. What is needed instead is for competent officers to be stationed in the field at all levels—from sergeant to general—who can operate on small amounts of foggy information and overcome friction at a quicker speed than can the enemy.

There is a parallel between planning and overseeing such operations and innovation portfolio management. It would be wonderful if every innovation project were staffed by brilliant project managers who had a perfect understanding of the market landscape including market needs, their own firm's capabilities, other projects that their firm is engaged in, the exact status of their projects, as well as their competitors' plans. As is clear to anyone who has engaged in innovation management, this is rarely possible. However, as we discuss, it is often possible for competent managers,

---

[1] For the first number of Abstraction-Creation, Art Non-Figuratif in 1932, Calder prepared a statement to accompany a reproduction of Little Universe (ontologicalmuseum.org/bookgiftshop/calder/calderbooks.html).

[2] Van Creveld, M.: Command in War. Harvard University Press (1987).

E.G. Anderson and N.R. Joglekar, *The Innovation Butterfly*,
Understanding Complex Systems, DOI 10.1007/978-1-4614-3131-2_5,
© NECSI Cambridge/Massachusetts 2012

acting on available information, to make good decisions in a timely manner if they use an appropriate management methodology.

Modern maneuver warfare breaks down planning and oversight of operations into a repeatable sequence of actions that is called the Observe–Orient–Decide–Act (OODA) cycle in order to cope with fog and friction.[3] The OODA cycle incorporates the idea of speed, i.e., the combatant with the faster OODA cycle should, all other things being equal, have an advantage on the battlefield by enabling a more flexible response to rapidly changing conditions. One common way to achieve a fast OODA cycle is to push down authority to the lowest possible subformation level, i.e., to radically decentralize authority.

Taking a cue from the OODA cycle, we argue that the innovation leader's ability to make good decisions when faced with innovation butterflies depends on the speed and flexibility of herself and her subordinates. These are achieved by speeding up the tempo of decision cycles at all levels of management, e.g., from the high-level aggregate portfolio planners down to lower level team leaders (such as those charged with the oversight of bench work in electronics or biomedical development settings). Importantly, the goal of this increased tempo of cycle time in the world of innovation management is *to cope not only with competitors but also with the ever-changing nature of market tastes*—in other words, the escalation of expectations.[4]

To truly increase the speed of the decision cycle, experts in maneuver warfare argue that top-level leaders should try to delegate as much of the decisions onto lower levels of command as possible. The rationale is that the opportunity to clearly see through the "fog" and to identify small but meaningful events as they unfold and then quickly connect the dots to detect the trends of emergent behavior, akin to the innovation butterfly, is only available at the lowest level. This suggests that in the realm of innovation, before we consider how to speed up the product innovation decision cycle at the aggregate level of a portfolio of innovation projects, we need to consider how to observe and speed up the management of individual tasks within a single project.

As all innovation leaders know, speeding up individual tasks in a project is not a trivial exercise. It should come as no shock that innovation projects, perhaps more

---

[3] The OODA concept was developed by the military strategist Col. John Boyd USAF in a series of briefings. Unfortunately, he never wrote an article or book, but many of these briefings can be found on the web. His ideas have also been compiled by Frans Osinga, Science Strategy and War, The Strategic Theory of John Boyd. Routledge, Abingdon.

[4] This decision cycle, referred to as Observe–Orient–Decide–Act (OODA) cycle in the military circles (see Coram 2002) is clearly parallel to the Plan–Do–Check–Act (PDCA) cycle popularized by the business guru W. Edwards Deming who helped Japanese industry to advance their design and manufacturing systems following WWII, and its successor, Define–Measure–Assess–Implement–Control (DMAIC) cycle from 6-sigma movement. However, OODA stands out in the sense that its architect John Boyd stresses the need to increase the speed of the decision cycle so as to get inside that of the opponent's (or opponents') in order to shape the military landscape.

- Edwards, D.W.: Out of the Crisis. MIT Center for Advanced Engineering Study. ISBN 0-911379-01-0 (1986).
- Robert, C.: Boyd: The Fighter Pilot Who Changed the Art of War. Little Brown, New York. ISBN 0-316-88146-5 and ISBN 0-316-79688-3 (2002).
- De Feo, J.A., Barnard, W.: JURAN Institute's Six Sigma Breakthrough and Beyond - Quality Performance Breakthrough Methods. Tata McGraw-Hill Publishing Company Limited. ISBN 0-07-059881-9 (2005).

often than conventional projects, come in late or over-budget in a majority of settings, *even if one excludes projects that are cancelled*. So what is the innovation leader to do? Fortunately, a close conceptual relative of the OODA cycle exists in the realm of innovation management in the methodologies of the so-called *agile development* movement that exists in software engineering realm. Agile development refers to a group of methodologies based on iterative actions, in which requirements and solutions evolve through collaboration between self-organizing cross-functional teams that review project prototypes against requirements as frequently as possible to keep the overall effort on target to satisfy the customer. This is a particularly useful method when customer requirements are ambiguous up-front, as is often the case in innovation settings. If a new product or service being developed involves primarily software features, then the translation of agile ideas into the product development realm is a relatively straightforward exercise. Admittedly, the application of these practices to other products, such as biomedical devices or the automotive sector is not common, primarily because of the difficulty of developing intermediate prototypes. However, with the advent of modern simulation technology, this particular barrier is beginning to become less problematic.

In contrast, the application of the agile methodology to manage multiple innovation projects, however, is poorly understood and much less common. Hence, we begin by describing the agile management of individual projects and then scale up to the agile management of portfolios.

The Agility Manifesto[5]

The field of software development has been notorious since its start (Brooks 1975) for delivering projects either late, over-budget, or with bugs—or more commonly, all three. Typically, a software program is initially architected as a sequence of logical steps.

---

[5] For a discussion of delays in software development see Brooks (1975). For a review of agile literature, see Abrahamson et al. (2002) and Tignor (2009). The term agility itself is not new in the mainstream new product development (NPD) domain (Thomke and Reinertson, CMR 2001). However, such terminology in the NPD literature refers to *inherent* uncertainty, rather than *emergent* uncertainty. Common prescriptions for improving performance in this literature include promoting flexible architectures and shorter response times. These ideas are congruent with agile development in the software engineering domain. However, for ease of exposition, we ignore the flexibility-based NPD literature and refer to *agile* practices as they reflect the current best practices in the software engineering domain.

- Abrahamsson, P., Salo, O., Ronkainen, J., Warsta, J.: Agile Software Development Methods: Review and Analysis. VTT Publications, Oulu (2002).
- Brooks, F.: The Mythical Man-month. Anniversary ed. In 1995 by Addison-Wesley Longman, Boston (1975).
- Tignor, W.: Agile Project Management, ISDC Conference, Albuquerque, NM (2009).
- Schweber, K., Beedle, M.: Agile Software Development With Scrum, Prentice Hall (2001).
- Thomke, S., Reinertsen, D.: Agile product development: Managing development flexibility in uncertain environments. Calif. Manage. Rev. (1998).
- The WIPRO example is taken from http://qualityconsulting.wipro.com/qualitycasestudy16.php.

These steps differ from hardware products because there are no physical laws (e.g., conservation of mass) involved in the development of software architecture. This gives software architects a great deal of flexibility in structuring their product, i.e., the software code, but it also makes them particularly vulnerable to changes in requirements for the targeted product features. The customers for software are aware of this flexibility, and unless there is a protocol or signed contract, they often change requirements late into the development process. If allowed, even small changes in requirements or in system architecture necessary to achieve them can easily lead to a cascade of changes resulting in a full-fledged tsunami of problems on developers' hands. Therefore, software settings are ideally suited for observing "innovation butterflies" and the propagation of their effects into the completed product.

Traditionally, software development projects followed a paradigm called "waterfall," wherein developed code follows a sequence of stages, each of which had a set of relatively infrequent reviews, called gates. The development of software by this process thus appeared to flow through these gates like a cascade of waterfalls on a mountain. Typically, the various components of the software project were developed mostly separately in the earlier phases of the product and then integrated together only during the latter phases of the project. The separation of effort and infrequent reviews led to a situation in which one change early in the process can set off a cascade of changes and rework during the later, integration phase. This put a premium on freezing customer requirements—even if the customer did not really know what they wanted—as early as possible during the process, which often led to less than stellar customer satisfaction. Even so, enough changes would usually accumulate so that projects spent an interminable amount of time in the integration stage of development.

A group of veteran developers in software industry recognized some of the problems with the traditional waterfall development process and came together in 2001 to issue a landmark document that came to be known as the *Agility Manifesto*.[6] The proposed agile development methodology welcomed customers to change requirements, even late in development, offered shorter timescales, and recommended frequent face-to-face communication sessions. It laid down a set of principles that argued that the best software requirements, architectures, and designs emerge from self-organizing teams, which are able to draw upon the needed expertise on the fly as requirements are changed.

Importantly, such characteristics make agile development attractive for innovation project management as well, because it also focuses on the effective and efficient management of gathering latent customer needs and converting these into feasible designs. In contrast, traditional software management—and for that matter traditional innovation management—processes are set up as a "waterfall" process similar to a manufacturing process such as the assembly of automobiles. However, innovation project management differs from manufacturing, in the sense that if a stage of manufacturing process is correctly executed for a given individual product,

---

[6] Taken from http://agilemanifesto.org/.

it need not be revisited.[7] Furthermore, most of the uncertainty in manufacturing stems from the variation in demand and supply volumes, which are simpler to forecast and manage than are the latent customer requirements for the functionality of a product. Unsurprisingly, because the agility manifesto was concerned with customer requirements, its proponents assumed that innovation management would be more amenable to the agile software development methodology than to the manufacturing-like "waterfall" processes that the software industry had used for the last 20 years (Beedle and Schweber 2001). We agree and so present a short primer on the agile development methodology because of its crucial role as a building block for maneuver-driven competition.

Agility Primer[8]

Three simple concepts lie at the heart of agile organization of software projects: backlog, sprints, and scrums, which we describe in the next sections.

Backlog: Based on product line, product, and system, an organization identifies all outstanding work and prioritizes it. This prioritized backlog list changes continuously, and is updated and reprioritized continuously.

Sprints: Like in running sprints, sprints in agile settings are short work increments in which a team works on completing an identified, self-contained group of prioritized backlog items, which constitute the work for that sprint. During the sprint, the backlog items worked on cannot be changed by a stakeholder who is outside the sprint team; although as work occurs within the team, additional work may be uncovered that may be dealt with either within the sprint or added to the backlog list for future sprints.

Scrums: The term is taken from the game of rugby where in players come together frequently in a huddle to move the ball forward.[9] Development teams gather in daily meetings to identify tasks that have been recently completed, tasks that must be done next, and possible impediments to those tasks.

---

[7] There are other foundational differences between innovation and manufacturing, e.g., manufacturing could not begin without physical inputs, but innovation can. See:

- Browning, T., Fricke, E., Negele, H.: Key concepts in modeling product development processes. Syst. Eng. **9**(2), 104–128 (2006).

[8] We have drawn extensively from many web blogs and discussions while developing this section. While we provide a few citations, it was difficult to cite all the literature, much of which is web based. Hence, we encourage the reader to access this literature. This particular section draws heavily from www.controlchaos.com. These terms have their origins in the game of rugby, where teams go back and forth while passing the ball.

[9] There is a parallel here in the new product development research context. Professors Hirotaka Takeuchi and Ikujiro Nonaka (1986) described a holistic approach to increase the speed and flexibility in product development and compared this dual approach to rugby, where the whole team "tries to go to the distance as a unit, passing the ball back and forth."

- Takeuchi, H., Nonaka, I.: The new product development game. Harv. Bus. Rev. (1986).

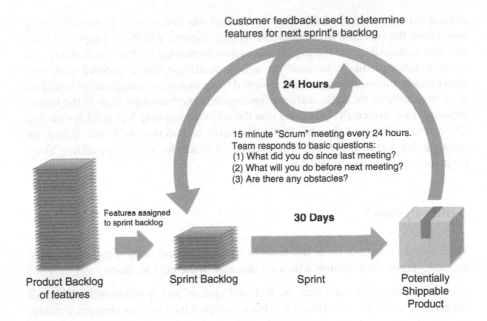

Fig. 5.1 Scrum flow diagram

As shown in Fig. 5.1, the scrum is at the heart of a process that can manage and control development work in an iterative fashion in an environment in which requirements are rapidly changing. It controls conflicting interests and looks for ways to improve communications and maximize cooperation. The scrum is also an adaptive way to quickly *detect and remove* anything that gets in the way of developing and delivering products. Agile developers believe that their methodology for organizing tasks is scalable from single projects to entire organizations, although how to do so remains only vaguely understood in settings beyond the software industry. (We discuss one possibility later in the chapter.)

Scrum-controlled and organized development and implementation has been used in the software industry for a decade now. It has also been scaled up to control the development of multiple interrelated products and projects, with large projects involving over a thousand software developers and implementers scattered over multiple firms. Practitioner and academic literature on this topic has been growing at a rapid pace. This literature growth reflects the maturity, and the success, of agile practices. Anecdotal evidence based on conversations with developers suggests that the incidents of butterflies begetting undesirable butterfly effects that may derail a project seem to be fewer in agile settings, when compared with traditional software development (e.g., WIPRO, a global software services firm, documents more than 90% reduction in errors, 65% reduction in manual effort, and 50% reduction in the turnaround time using agile development practices when compared with the traditional waterfall methodology).

Aside from pure software projects, scrum processes have been used to produce financial, Internet, and medical products in which software is a product offering.[10] However, applying adaptive techniques (such as scrum) beyond the pure software domain, to a combination of hardware and software development, or in settings that cut across a firm's boundary, creates its own coordination challenges. With respect to hardware, physical laws come in play making it difficult to remove or ignore certain tasks as the result of a scrum meeting or a sprint review. In contrast, with multiple firms, the principle of distributed providence suggests that coordinating the actions of individual team members from different firms (or even the same firm, but with teams in different time zones) may prove problematic.

Case: Video Game Development

We now present case evidence on such challenges from the video gaming industry. Readers may recall the industry dynamics and competitive nature of the video game industry described in the introductory section. We now look at management practices used in individual development projects within this industry. Figure 5.2 shows the evolution of a project development path against an imagined "best" path as this project traverses through various phases of development.

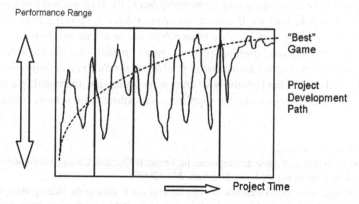

Fig. 5.2 Game performance vs. time in agile development. Adapted from Sellers, M.: The stages of game development. In: Francois Dominic Laramee (ed.) Secrets of the Game Business. Charles River Media, Hingham, MA (2005)

---

[10] Cordeiro, L., Mar, C., Valentin, E., Cruz, F., Patrick, D., Barreto, R., Lucena, V.: An agile development methodology applied to embedded control software under stringent hardware constraints. ACM SIGSOFT Software Engineering Notes January 2008, vol. 33, No. 1 (2008).

This figure is taken from the work of Michael Sellers, CEO of a video gaming software firm, Online Alchemy.[11] He argues that in some extreme cases *"you might find that you are not converging on a single game. This can happen if you did not start with a clear concept, or if the external requirements change throughout the life of the project."* Each time a new requirement shows up, the developers have to work it into the next sprint through a cycle of "Define → Design → Implement → Evaluate" tasks to implement it. They may have to repeat this cycle if, after a sprint review involving the customer, the current project does not meet the requirement or if the requirement interacts with some portion of the game that falls short of its projected performance relative to customer expectations. For example, the color pallet may have 512 choices, but the rendering is slow, so the pallet is reduced to 256 choices. This solves one problem, but because of the principle of exchange, the parameter setting can set many butterflies in motion in order to meet the esthetic requirements of the customer. Existing video clips may have to be modified, characters may have to be redesigned, and color contrasts throughout the game adjusted.[12]

We asked Michael Sellers how his teams have managed such development projects. In his view, *"The "Waterfall" model won't work because you don't know the requirements up-front. The [agile] model works better, but we insert an extra step, not only to review the requirements but to ask: does this feel fun? Is this the type of game we want to be making? You need a design document in one sense. But on the other hand, a constellation of small documents can work, if you have one person keeping it in their head. And then you grow these documents in parallel through iteration. Then you go back into agile mode. This gives a lot [of flexibility] to the team, but you have to accept a lot [of changes] back. We'll chunk out tasks and subtasks for four weeks and we'll commit to always have something playable (also known as "never going dark") ... we have a backlog of all the things we want to do and then for this release—perhaps one or more sprints—we decide what we're going to do and convert this into a list of backlog tasks that need completing."*

We also asked Michael about how emergent issues are managed, particularly when development involves outside suppliers. According to him, most coordination

---

[11] Sellers, M.: The stages of game development. In: Francois Dominic Laramee (ed.) Secrets of the Game Business. Charles River Media, Hingham, MA (2005).

[12] This type of expansion and contraction of open items is not limited to the video gaming industry. Similar data have been reported in development processes at Microsoft (Cusumano and Shelby 1998). The second author has analyzed analogous challenges using a project completion status dataset derived from Ford Motor Company's automotive styling effort (Yassine et al. 2003). In the Ford Case, the external requirement changes were isolated, and it was shown that the design path oscillated (and hence we coined the term "design churn"), due to the interconnected nature of design tasks—i.e., based on *the principle of exchange* discussed in Part I—even when the butterflies were not created by outside interventions from customers.

- Cusumano, M., Shelby, R.: Microsoft secrets: how the world's most powerful software company creates technology (1998).
- Yassine, A., Joglekar, N., Braha, D., Eppinger, S., Whitney, D.: Information Hiding in Product Development: The Design Churn Effect, Research in Engineering Design (2003).

with suppliers is based upon specifying standards, e.g., of components such as graphics rendering, physics, or sound "engines." Misaligned goals and emergent conflicts between the game developed in-house and the supplier engines are inevitable. So his team uses "demos" showing in stripped-down visual or audio form what they expect the supplied "engines" to do to reduce the possibility of missing functionality or other problems. When problems do emerge, Michael's team has had to go out and buy additional software modules or services from third party suppliers or, sometimes, they just have to develop the solutions internally. Just as we described in our earlier discussion of scrums, Michael emphasized that they must account for multiple stakeholder perspectives. Using resources outside the agile team such as open sourcing unleashes butterflies and imposes unique constraints that the team has to work around.

To summarize, *even agile is supposed to be agile,* i.e., the innovation leader must be prepared for the fact that emergent technologies or requirements may force the team to adjust standard agile development practices on the fly. Next, we review the application of agile ideas to management of portfolios of projects.

Extending Agile Processes to Multiproject Development

We now draw upon theories of *pattern recognition and scaling* from complexity science and maneuver warfare from military science to translate agile development principles from projects to portfolios. Recall that a central lesson in complexity science is that complexity could be reduced in terms of pattern recognition by moving to a higher level of aggregation in order to examine patterns. Thus, a central lesson from scaling is that *relevant measurements must change as one goes up and down the ladder of abstraction.* In the realm of innovation, while managing a project, a manager usually focuses on the interactions between tasks and any resultant rework, as indeed he should. On the other hand, while managing a portfolio, higher-level management should instead focus on the interaction between projects and their impact on the market place in aggregate just as scientists and engineers ignore the movements of individual gas molecules in order to focus on the gas's aggregate temperature, pressure, and volume.

Parallel problems have long been studied by the theorists and practitioners of maneuver warfare. We adapt and apply their solutions to expand the concept of agile management to embrace portfolios of innovation projects and capabilities. The result, which we call "maneuver-driven competition," expressly copes with the uncertainty created by the innovation butterfly by making rapid adjustments at the portfolio level.

As discussed earlier, uncertainty in military science is captured by the twin concepts of "friction" and "fog." The cornerstone for dealing with both types of uncertainty is to break down planning, oversight, and execution into a rapidly repeatable sequence of actions that is called the OODA cycle. The agile methodology nicely mirrors this in the realm of innovation at the project level, coping well with the twin problems of "fog" and "friction" that beset innovation as well as warfare. Unfortunately, while much has been written on the subject of agility at the project level (see note 4 in this chapter), comprehensive schemes for implementing agility ideas at the portfolio level is lacking. However, the maneuver warfare literature includes certain principles that we can use to inform the thinking of agile innovation portfolio leadership. The first principle is that there ought to be some uniformity of organizational culture and procedures, such as the backlogs, scrum meetings, and sprints from the agile methodology. These principles need to be expanded upon during the aggregate planning of capabilities (i.e., development resources), which we do in Chaps. 6 and 7. For now, we will assume that such uniformity in culture and procedures exist. A key idea for upper management to consider during the implementation of agile portfolios is what some military science theorists call a "directed telescope"—in management parlance, a "managerial representative"—in the field. Beginning in the 1850s, armies that practiced maneuver warfare began to regularly send staff officers to report tactical information from the field up to the higher commands and relieve some of the burden of interpreting and reporting trends from the local commanding officers in the field, who had too much to do anyway. The data received through these mechanisms also provided an independent viewpoint, different from the data reported by the field commander. A higher-ranking commander, away from the field of battle, could then synthesize this information and use it to orchestrate the actions of the various field commanders and their teams. Similarly in agile-managed projects, a member of the scrum team—typically, a management representative—reports the progress to a higher-level innovation leader (which we shall term the *innovation executive* to separate this leader from the innovation project manager) and conveys the innovation executive's intent for the project back to the innovation project manager. The key operational question is how best to use the managerial representative and uniform doctrine to improve the development and usage of capabilities and thereby improve the performance of the product portfolio? We now introduce the operational ideas of scouting, roadmapping, orchestration, and maneuver to illustrate the key elements of a cycle for agile portfolio management.

Agile Portfolio Management Cycle: SROM

The Scout–Roadmap–Orchestrate–Maneuver (or SROM) cycle for innovation management is presented graphically in Fig. 5.3. Each phase of the cycle is described in detail thereafter.

Scout

At the level of the innovation portfolio manager, who controls multiple, interlinked projects, the first order of business is to understand the market landscape as well as

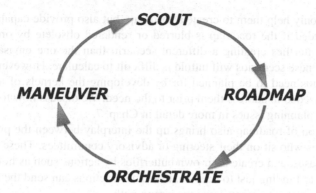

Fig. 5.3 The SROM cycle of innovation portfolio management

the portfolio of products and capabilities—and those of the firm's competitors—in order to make new products. Of course, the market landscape—just like the military landscape—is foggy but a foggy picture is better than none at all. To use a sports analogy, most good sports teams scout their competition by watching films of previous games. Watching these films does not guarantee visibility into a specific game plan that the competition will deploy, but it can at least reveal player tendencies and team patterns.

One of the most difficult parts of scouting is figuring out what it is that one looking for. The senior managers' ability to "scale" the problem and communicate only the key parameters to the lower level managers plays an important role here. If these parameters are communicated crisply, people closest to the detailed phenomena can identify key patterns as they emerge.

Roadmap

The next task for the innovation portfolio manager is to figure out a technology roadmap that will benefit the company and make sense within the context of the firm's capabilities and culture. Typically, this will require the development of series of sequenced product introduction and hiring decisions to build the firm's capabilities while creating these products. Both of these decisions; product introduction and hiring must be designed so as to create capabilities that provide a competitive advantage. For example, early on, when the Ford Motor Company lacked the resources to design appropriate electronic engine controls, it decided to create a capability in engine control electronics by (1) training some former carburetor engineers to become electrical control engineers while, at the same time and (2) hiring literally hundreds of freshly trained electrical engineers during the early 1980s. However, sequencing product introduction and hiring decisions for the current roadmap is not enough. First of all, the roadmap's planners must account for the fact that all projects will face an unknown amount of friction and plan accordingly. Even more crucially, innovation executives and their staffs must also determine a set of capabilities that will not only help them to create a roadmap, but also provide capabilities that might be needed if the roadmap is blurred or rendered obsolete by one or more innovation butterflies creating a different scenario than the one envisaged. How likely any of these scenarios will unfold is difficult to calculate. However, the more likely scenarios need to be planned for by developing the kernels of appropriate capabilities needed to address them prior to the need. We discuss how to cope with both of these planning issues in more detail in Chap. 7.

The creation of roadmap also brings up the interplay between the project team and executives who sit on their steering or advisory committees. These executives often have biases, and create their own butterflies by actions such as holding on to purse strings to funding just to keep control. These actions can send the wrong signals and shape the road maps down the wrong path.

Orchestrate

Both the scouting and road-mapping activities should be going on all the time. Furthermore, every so often, an innovation executive needs to initiate an assessment of how well the projects align with the roadmap by considering the status of a project and its purpose statement. These purpose statements should not only include information on what is to be done by a project, but also why. That is, the project statement should include a statement not only of project goals but also of the project's purpose for the firm as a whole. The innovation executive then needs to adjust and disseminate these statements to each of the project teams so as to ensure their coherent action.

Aside from reading the tea-leaves at the portfolio level, probably the most difficult parts of this cycle for the innovation executive is to step back and not micromanage the various project groups once they have been set off to work. Each individual project will run into its own butterflies. Some of these butterflies will become important and potentially "game changing"; some will not and merely create "friction" analogous to that experience in warfare. The temptation for the innovation executive to intervene can be quite high. It should be avoided because it can be harmful because the project managers "on the ground" will likely be better informed than the executive about the projects' sources of friction and ways to overcome them as well as project opportunities that are potential butterflies. The project manager can only do this effectively, however, if he understands the "why" of the project. Otherwise, he is likely to make a choice when coping with a butterfly that will create a misalignment between his team and the rest of the firm.

As a rough guide, the innovation executive should consider intervention only in one of three situations: (1) if a project manager or the project team's management representative identifies a potential butterfly that will impact other projects; (2) if a project goes seriously off-track in terms of its projected objectives; or (3) if the motives for that project suddenly change or become nontenable due to changes in the business environment, perhaps because of butterfly effects from other projects. Hence, the innovation executive should behave much more like a conductor of an orchestra than the traditional picture of a hands-on manager. The conductor does not tell individual flute players *how* to play the notes louder or faster, for example, only to play louder or at a quicker tempo. Similarly, the innovation executive must guide the individual project teams toward coherent action without involving herself in the details of any individual project.

Maneuver

The individual project groups must then go about their business of creating or updating the deliverables for their projects, guided by the project's goals and purpose. This is where the traditional agile project management methodology for individual projects (described earlier in this chapter) as a part of the agility manifesto is executed.

To some extent, the orchestrate and maneuver phases of the decision cycle overlap. Maneuvering begins once the project goals and purposes are distributed to the project managers from the innovation executive during the orchestration phase. However, like the conductor of an orchestra, the innovation executive does not stand idle during the maneuver phase. While the executive should not intervene directly in project execution except under dire circumstances, she still needs to have a general idea of what is going on within the project and how well it is proceeding. Is it making sufficient progress and on the roadmap? Is it deviating from its goals, and if so, why? Unlike an orchestra conductor listening to the musicians, however, an innovation executive cannot directly observe the progress of all projects. This is another major purpose of having a directed telescope in the form of a managerial representative on each scrum team—to provide a view into the project's progress for the innovation executive. By "being on the ground" of the project, the managerial representative can provide a particularly valuable service because his view of the project will be less obscured by fog and organizational distance than the innovation executive's. Importantly, the managerial representative does not supplant the project manager but supports him by creating an additional channel of information to the innovation executive and her staff.

The managerial representative, because he has direct knowledge and personal connections with bench level work, can also help with another problem. A weakness in the analogy between an innovation executive and an orchestra conductor is that the innovation roadmap is constantly changing and in some sense resembles improvisational jazz more than say, classical music, in which each note is fixed in the musical score. This then is the other major purpose of the managerial representative: to feed back information on the firm's capabilities development and any emergent innovation butterflies that could provide an opportunity for the firm to exploit. These observations are ultimately inputs to the next cycle of scouting and roadmapping.

Scouting Again

When there are long delays associated with completing projects and even longer ones in building competencies, friction and fog may end up requiring readjusting the roadmap before it is completed. Furthermore, the emergence of an innovation butterfly may force the roadmap to be entirely discarded. Hence, the SROM cycle should ideally result in only a short maneuver within the innovation space before a new scouting cycle begins, which is then followed by yet another cycle of road-mapping, orchestration, and maneuver. In general, a quick SROM cycle is desirable because it will enable a firm to (1) adjust more quickly than its competition to a changing environment and (2) give the firm a chance to exploit any innovation butterflies before they fly away. This leads us to use the term *maneuver-driven competition* to describe the SROM cycle because *only short maneuvers can be carried out before the uncertainties resulting from "friction," "fog," and—most impor- tantly—innovation butterflies make the existing roadmap obsolete.*

That said, this "need for speed" in the SROM cycle must be balanced with the fact that most disturbances will not push the system onto another path. They are, in fact, simply friction that can be overcome as the individual project level. However, these disturbances, combined with the various estimation biases as discussed in the *principle of distributed providence* (see Chap. 4), contribute to making scouting akin to surveying a foggy landscape, which even the best managerial representatives and project team's cannot completely dispel. Like an automobile driver who pru- dently slows down under foggy driving conditions, because it is difficult to see the actual road conditions up ahead of him, the innovation executive cannot iterate the SROM cycle too quickly. This is particularly true in the scouting phase because this is the phase in which data is collected, collated, and filtered prior to informing the planning process. This will necessarily take more time if the data is more ambigu- ous. If enough time is not taken to scout, the firm runs the risk of having its teams reacting to inaccurate information, thus becoming much like a dog chasing its own tail. This is analogous to the adage from W. Edwards Demming, the father of statis- tical process control: Avoid reacting too quickly to normal variation in a process with a process change, lest you inadvertently make that process even more variable than it would be if you had not tampered with it in the first place.[13]

A second time-related dimension of orchestration is the time required to pick up market cues, assess portfolio-wide risks and benefits, and then signal to the indi- vidual teams what types of adjustment may be desirable and why. How to cope with these two issues, so as to speed up the SROM cycle in a useful manner, is discussed as part of modularizing innovation portfolio risk, which is the subject of the next chapter.

---

[13] This adage has been attributed to W. Edward Deming—however, we have not been able to find a suitable citation. Managerial overreaction has also been known to have a "tampering effect." This effect can sharply reduce the effectiveness of the SROM cycle.

Chapter 6
Modularizing Portfolio Risk

Install Firewalls to Prevent Runaway Effects

> *Well, when you're trying to create things that are new, you have*
> *to be prepared to be on the edge of risk.*
>
> Michael Eisner[1]

Management of innovation risks is central issue on the minds of new product development managers.[2] A reader may recall from the innovation system as presented in Chap. 2 that the delays involved with developing capabilities and delays in executing projects that use these capabilities are major drivers in the complexity of the innovation system. The fact that innovation workers build their capabilities by executing projects only complicates this picture. In short, a key source of complexity of the innovation system is due to the feedback loops that exist between these capabilities and the associated projects required to develop them. While much complexity (and potential for creating harmful innovation butterflies) will remain no matter what the innovation executive and her staff do, things will improve if roadmaps can be made more robust; that is, if the planners can somehow reduce the chances of a roadmap going awry and requiring adjustment or wholesale redesign. If this can be done, the predictability of the innovation system—at least in the short run—will improve, thus reducing the "fog" inherent in innovation systems and enabling the firm to execute the Scout–Roadmap–Orchestrate–Maneuver (SROM) cycle—a crucial underpinning of maneuver-driven competition—at a quicker tempo.

---

[1] http://www.woopidoo.com/business_quotes/authors/michael-eisner-quotes.htm.

[2] For a review, see: Sommer, S., Loch, C., Pich, M.: Project risk management in new product development, Chapter 17. In: Loch, C., Kavadias, S. (eds.) Handbook of New Product Development Management. Butterworth-Heineman, Oxford (2008).

E.G. Anderson and N.R. Joglekar, *The Innovation Butterfly*,
Understanding Complex Systems, DOI 10.1007/978-1-4614-3131-2_6,
© NECSI Cambridge/Massachusetts 2012

In order to understand how best to do this, we begin with the following example of a firm coping with a major innovation butterfly by *modularizing* the risks in its innovation portfolio by reducing the interaction of individual project and capability risks at the task level. Hence, if task A in a particular project is delayed, it will not affect task B in some other project, and vice versa.

Case: Technology Roadmap at World Motors

One of the key jobs of any R&D organization is to manage the interactions of the capabilities of its workforce, both internal and partners, across a suite of ongoing projects. To begin to understand how these capabilities might interact, consider the case from World Motors (WM) below, which is currently geared up to meet a projected legislative shock in terms of fuel consumption requirements.[3] This legislation is expected to require an increase in the average fuel efficiency of all motor vehicles sold in the market segments where World Motors is a participant. WM's automotive engineers have been planning for this scenario in two ways. One is for the auto body engineers to learn how to use their new computer-aided design (CAD) system to better represent the position of each component in three dimensions (3D), which will enable the engineers to pack auto components more tightly and ensure that they fit together properly in the actual prototype. This will also enable the engineers to reduce the size of the engine and other auto components, which translate to a smaller body, lower weight, and increased gas mileage. Every pound saved will result in increased fuel efficiency without raising parts costs. WM will launch this CAD design tool on a roadster (a two-seat sports car with an open top) to be introduced in the 2011 model year.

However, weight savings will not be enough to increase gasoline efficiency. WM will also need to develop new engines in which each cylinder is monitored and controlled by a microprocessor to increase fuel efficiency. To do this will also require the development of a newly designed engine controller that integrates inputs and outputs generated by the microprocessors for each cylinder. Wanting to avoid developing highly interactive technologies that require radical changes simultaneously, World Motors decided to pilot the new controller on a small minivan to be launched in 2010 and the new engine on the previously mentioned two-door roadster to be introduced in 2011. The year 2012 will be a quiet year in the sense that no new technologies are planned for pilot projects.

Each of the three vehicles to be launched from 2010 through 2012 requires four major tasks: (1) development/adaptation of the controller, (2) development/adaptation of the engine, (3) body development, and (4) integration of the auto body, controller, and engine. Because the same type of engineering specialists are needed for each task for each vehicle (i.e., the body engineers who work on the minivan will be the same engineers who work on the roadster and subsequently the 2012 sedan

---

[3] This is an assumed name and a stylized case. Key facts are drawn from existing evidence, but the details have been scaled, and some additional material has been added, to make this a stylized case.

project), scheduling buffers have been included in each vehicle's project team in case things go wrong, which they always do!

For example, additional time has been planned in the initial development of the minivan's controller. Part of this is because a new technology requires one-off tasks, as semiconductor vendor selection during the microprocessor design, that for the most part will not need to be repeated such. Part of that time, however, will be used to train the engineers on how work with the selected vendor to design a parallel processor-based controller; in other words, to develop their capabilities as engineers.

Similarly, the body development phase of the roadster, which is 3 months longer than the corresponding development phases for the launch of the redesigned minivan and sedan. The reason for this is that the body engineers need to develop the capability to specify components on the computer as three-dimensional (3D) solids using the new CAD system mentioned above, whereas in the old software they had previously used deployed relatively difficult to visual mockup made from two-dimensional (2D) representation of the parts. Hence, the 3 months corresponds to the time needed for the body engineers to develop a new 3D visualization capability. Once these capabilities are developed, however, they do not need to be redeveloped for the next product, i.e., the 2012 sedan project. Hence, body development reverts back to its original 9-month schedule, and the integration time budgeted for the sedan project is less than that needed for its predecessor as well.

The Paralyzing Curse of the Project Domino Effect

If these CAD capabilities and components are not developed on schedule, the consequences for WM could be severe. In particular, its sales of roadsters, which are high-profit-margin vehicles but also less fuel efficient, may be restricted. Even if the roadster is completed on schedule with the new engine, controller, and body design technologies, if the sedan is late, its roadster sales will be affected because the proposed legislation restricts the average fuel economy of *all* vehicle sales for the entire model year. That is, in order to sell each the high fuel consumption roadster, WM must also sell a low fuel consumer vehicle, i.e., the sedan. Thus, every sedan that is not sold during the 2012 model year because of a late introduction, one less roadster can be sold, which creates a double loss in overall profits.

WM has built in 3-month buffers as delay contingencies in each of the three projects to prevent such an outcome as shown in Fig. 6.1. However, these buffers may not suffice if certain events occur. For example, if body development for the roadster proves more problematic than expected because of bugs in the new software (WM is the first automotive user of this new software), then WM might have to use the old software to develop the roadster. This could translate not only to a poor outcome for the roadster, which will restrict its sales in the 2012 model year, but it also suggests that the body development engineers will have to complete developing their new 3D solid design capability during the sedan project, which is problematic because extra time for capability development has not been budgeted. Thus, a delay in the 2011 roadster's development can cause delays in the development of the 2012 sedan, as well.

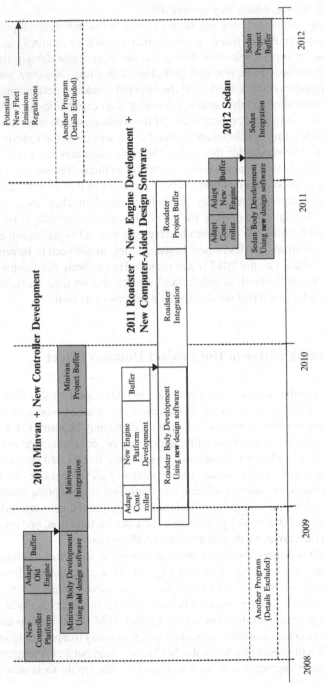

Fig. 6.1 Development roadmap at World Motors

This domino effect becomes even more serious if the pilot production (also known as tape-out) for the minivan's engine controller chip is flawed, holding up engine-controller design completion for several months. This will present two threats to the projects' schedule. The incorrect tape-out could push back the start and completion dates of the minivan's integration phase, which will not only affect the minivan's ultimate time to market because the same engineers are used for integration of the roadster, but it will also likely delay commencement of roadster integration as well. A second problem is that the new controller must be completed with the 2010 minivan for development of the new engine for the roadster in 2011. Hence, if the pilot production of the 2010 controller is flawed, then an old controller must be adapted and used so that the 2010 minivan might be launched on time. This decision will leave the 2011 roadster program in the lurch because it needs a new engine controller and its schedule only allows enough time for the adaptation, but not the development of a new controller. Further, the 2012 sedan requires that both the controller and engine will be adapted, not developed. Hence, if any phase of the roadster's timing is delayed—engine, body, or controller development, and/or integration—the schedule for the 2012 sedan will be delayed as well. That is, because of the interactions of WM's development and reuse capabilities across models, a single delay in the controller development in 2010 could be the first in a series of dominos that topple each other down and could result in huge market losses.

Capability Interactions

Modern product architecture theories espouse practices that restrict interactions, such as those caused by information exchange, heat flow, size, and shape, between components.[4] Such restriction of the various interactions between components is

---

[4] For deeper discussions of architectural choices, see:

- Ulrich, K.: The role of product architecture in the manufacturing firm, research policy (1995).
- Baldwin, C.Y., Clark, K.B.: Design Rules. The Power of Modularity. Vol. 1. MIT Press, Cambridge, MA (2000).
- Ramdas, K.: Managing product variety: an integrative review and research directions. Prod. Oper. Manage. 12(1), 79–101 (2003).
- Yassine, A., Wissmann, L.A.: The implications of product architecture on the firm. Syst. Eng. 10(2), 118–137 (2007).
- Krishnan, V., Ramachandran, K.: Economic models of product family design and development, Chapter 4. In: Loch, C., Kavadias, S. (eds.) Handbook of New Product Development Management. Butterworth-Heineman, Oxford (2008).
- Ro, Y., Fixson, S., Liker, J.: Modularity and supplier involvement in product development, Chapter 9. In: Loch, C., Kavadias, S. (eds.) Handbook of New Product Development Management. Butterworth-Heineman, Oxford (2008).
- Gomes, P., Joglekar, N.: Linking modularity with problem solving and coordination efforts. Manage Decis Econ 29(5), 443–457 (2008).

commonly called modularization. One way to achieve such modularity, in order to gain either production or usage efficiencies, is by specifying interface requirements, so that various components can be developed or services in standalone manner, yet they fit nicely when integrated. If the interface is agreed on by a number of firms, it is often referred to as an industry standard.

However, interactions can also occur among a firm's development capabilities, although the interactions are not always as obvious as merely looking at the interface specifications. Looking at GM roadmap in Fig. 6.1, we see a number of such interactions, some of which might cause a project domino effect under certain conditions. An example of this is if the engine controller engineers are preoccupied with a project delay in minivans when they are needed to work on the roadster, they cannot work on both projects simultaneously without delaying both or cutting corners in terms of their deliverables. Many of these interactions are referred to by project managers as "resource conflicts," in which the professionals needed for one project are currently engaged on another project.[5]

Because human beings are the source of a firm's capabilities, any cross-project interaction of resources can result in a capability dependency. For instance, a problem could occur if WM's engine controller for the minivan project is outsourced. Then WM's internal engineers will not have garnered enough experience working on individual engine cylinder controllers to complete the engine controller for the roadster in the planned amount of time. In essence, the controller engineers will not have mastered the "experience curve" of developing a capability to develop engine controllers (because another outside, outsourced group did this work) and hence will require more time to develop the controller for the roadster. A similar effect will impact the sedan's timing if the body engineers do not climb the "experience curve" of 3D solid modeling specification during the roadster's development. In principle, some of this learning could be used across projects by the transfer of documentation, intensive meetings, and/or formal reviews. However, these tasks are not trivial. Documentation takes[6] time to develop and usually is incomplete, lacking many points of tacit or "tribal" knowledge. The best way to transfer these types of skills from one project to another is through the transfer of people from the first project to the next where the know-how could be applicable.

Specification of interfaces for individual projects and conventional roadmaps that indicate plans to reuse components from one project to another do not address these people and know-how transfer issues. Perhaps, it is possible for some engineers

---

[5] We refer the readers to the work of Professor Nelson Repenning, and his collaborators, on such resource conflicts and their behavioral implications owing to what he terms as the "Firefighting" effect.

- Repenning, N., Gonclaves, P., Black, L.: Past the tipping point: the. persistence of firefighting in new product development. Calif. Manage. Rev. (2001).

[6] Professors Nonaka and Takeuchi were among the first set of scholars to examine the tacit nature of knowledge in new product development settings:

- Nonaka, I., Takeuchi H.: The knowledge creating company: how Japanese companies create the dynamics of innovation. Oxford University Press, New York (1995).

to do double duty and devote a part of their time to the follow-on project, even if the earlier project is delayed. However, as with any partial transfer of people between projects—for example, the transfer of a core group of body engineers from the roadster prior to its completion to the sedan's body development—will tend to slow down the earlier project. Nor will this transfer speed up the later sedan project as quickly as would the transfer of the bulk of the engineers to the sedan body development upon the completion of the earlier roadster project.

Accounting for Disparate Types of Capabilities

Some capabilities may also be dependent upon the prior development of other capabilities within the firm. One of the assumptions that World Motors is making in developing its new, more fuel-efficient engine for its roadster is its prior development of the parallel processor engine controller for its minivan. If the controller capability is not developed for the minivan, and it remains unavailable for the roadster, then developing a new engine to fully take advantage of the speed of the parallel processors in controlling each engine cylinder is irrelevant. We refer to this as *technology roadmap dependency* between two unique capabilities.

A related type of dependency can occur at an even more fundamental level. According to Fine and Whitney (1996), Toyota builds a significant percentage of its transmissions itself even though it would be cheaper to have a supplier build them.[7] However, Toyota's management feels that without actually building at least some of the transmissions itself—that is, without building a transmission development capability—the company will be unable to design their engines and other components so as to get the best performance out of the transmission and, by extension, the car as a whole. That is, some capabilities are simply not strong without the presence of other complementary capabilities within the firm. Probably, the capability that is most likely to create butterflies is the ability of a team to "integrate" a product. That is, to decompose the product into modules for the detailed development of parts and then weave or reintegrate the completed modules back into a coherent defect-free product that satisfies the customer. The World Motors case reflects this in that it achieved a shorter integration period for its roadster once it developed complementary capabilities of developing parallel-processor-based controllers (by its control engineers) and 3D solid specification (by its body engineers).

Another way to build capabilities is through the usage of the markets. A classic example of this approach is the iTunes Store online music offering. This offering itself is a triumph in simplicity for downloading online music in a near-turnkey manner. However, it is hard to imagine its current success without the earlier development of the iPod, which featured an intuitive user-interface that revolutionized—and in many ways defined—the digital music player industry. Together, the two

---

[7] Fine, C., Whitney, D.: Is the make-buy decision process a core competence? MIT CTPID Report (1996).

Table 6.1 Various types of capability interactions

Dependency types[a]	Examples
Resource dependency	World Motors Minivan and Roadster Controllers Timing
Roadmap dependency	World Motors Minivan Parallel Processor Controller and Roadster Engine
Complementary dependency	Toyota transmissions and engines, World Motors sedan integration
Market dependency	iTunes Store and iPod

[a]For a fuller discussion of various types of dependencies, see:
- Malone, T., Crowston, K., Lee, J., Pentland, B., Dellarocas, C., Wyner, G., Quimby, J., Osborn, C., Bernstein, A., Herman, G., Klein, M., O'Donnell, E.: Tools for inventing organizations: toward a handbook of organizational processes. Manage. Sci. **45**(3), 425–443 (1999)
- For an application of dependency ideas, see Balasubramanian, P.R., Wyner, G.W., Joglekar, N.R.: The role of coordination and architecture in supporting asp business models. Proceedings of the Hawaii International Conference on System Sciences (HICSS-35) (2002)

capabilities, iPod plus iTunes Store, provided a turnkey solution for the download of digital music from the Internet into a player/recorder that could be used to play music in a convenient, easy-to-use manner anywhere at all.

Table 6.1 presents various types of interactions between capabilities. An important point to consider when examining the interactions between capabilities is not only how the different capabilities interact during a single project, or across multiple simultaneous projects, but how they *interact with each other over time*.

Managing Capabilities Under Emergence

How does the principle of escalation of expectations and the principle of exchange affect the interactions among capabilities? As described in the World Motors case, because of the interdependencies between capabilities over time, a number of risks to what might appear to only one project will in fact affect multiple projects. The innovation system is highly path-dependent because of the escalation of expectations. Furthermore, managing this path dependence is extremely difficult even if only a few disruptions occur upfront because of the principle of exchange. How can innovation managers gain some control over the system in order to effectively perform overall product and capability portfolio planning over a long time horizon?

Some product development researchers have suggested that the key to understanding product development is to consider product development as a *"journey through a rugged landscape"* such as mountains as shown in Fig. 6.2.[8] This is a very useful metaphor for some purposes, particularly when looking to improve a product's performance

---

[8] Loch, C., Kavadias, S.: Managing new product development: an evolutionary framework, Chapter 1. In: Handbook of New Product Development Management. Butterworth–Heineman, Oxford (2008).

Also, Chapter 6 of this handbook provides a comprehensive discussion of search over a complex (but time invariant) landscape during product portfolio management: Kavadias, S., Chao, R.: Resource allocation and new product development portfolio management (2008).

Fig. 6.2 Innovation as search on a rugged landscape.[9]

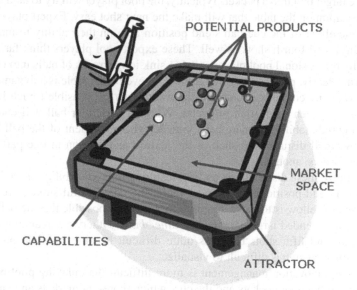

Fig. 6.3 Portfolio management as pool play.[10]

incrementally, because there are limits to how far you can push a product's performance (e.g., increasing the speed of a horse-drawn buggy) before one has to metaphorically climb down one mountain and begin to ascend another (perhaps by substituting the horse-drawn carriage by a car powered by the internal combustion engine).

However, for simultaneous product and capability portfolio planning, a new metaphor may be more vivid and helpful. Consider instead the portfolio planning problem as a game of pool as shown in Fig. 6.3.

---

[9] Figure created with Microsoft PowerPoint and used with the permission of Microsoft.
[10] Figure created with Microsoft PowerPoint and used with the permission of Microsoft.

In this case, the table surface is the market and the billiard balls correspond to the products in a portfolio. Like portfolio managers, pool players can only interact with billiard balls ("products") through a pool cue stick and cue ball, which we compare to a capability. The player is trying to develop products into the appropriate product space, which corresponds to sinking each of the balls into their desired pockets. Like the innovation system, a pool table is a dynamically complex system capable of extreme path dependence, which frustrates many novice players to no end, but it also gives them numerous options for getting a ball into a pocket. Naïve and inexperienced players are lucky to sink one or two balls in a row. However, expert players can typically sink many balls into the appropriate pockets in a row before missing.

To accomplish such a sequence of success, most good pool players follow two cardinal rules which are exemplars for our view of maneuver-driven competition. The first is to consider where the cue ball will end up on the pool table after it knocks the target ball into a pocket. Typically, the pool player will try to land the cue ball in a location on the table that will make the next shot easy. Expert players will consider the effect of the cue ball's end position also on their ability to make the second, third, and fourth shots as well. These expert pool players think far ahead. This is why professional pool players tend to sink long streaks of balls into pockets in one turn. Clearly, playing the game of pool on even a flat table is a dynamic process, where many complex interactions between balls are possible (much like the interactions between innovation projects). Where a particular ball will end up is prone to unpredictability if many balls bounce. The movement of the ball is also affected by small disturbances such as the texture and friction in one part of the playing surface vs. another.

To minimize these disturbances and the resulting unpredictability of the billiard balls' moving end-positions on the pool table, the other useful strategy for good pool players to follow is to make sure that to the extent possible the cue ball, once it has sunk its intended target, *touches no other balls*. Otherwise, planning for the third, fourth, and fifth shots becomes quite difficult because multiple balls will move and the next step is difficult to visualize.

Of course, innovation management is more difficult than playing pool because the location of the pool pockets and the cue, which represent markets and capabilities, can change over time! Nevertheless, the rules of expert pool playing are still instructive. Like an expert billiard player, a skilled innovation project portfolio executive needs to first pick projects and align them to make the best shots. She then needs to (1) pick current projects so that future projects become "easy to sink into the pocket" through road mapping as discussed in the previous chapter and (2) ensure that these project choices minimize the chances that the firm's capabilities interact with each other in an unexpected manner, i.e., *the manager must modularize the risk to the firm's current and planned capabilities*.

Beyond Playing Pool: Back to the SROM Cycle

How does playing pool relate to the SROM cycle described in the previous chapter? Clearly, an innovation executive needs to "pick the right shots in the right order" during the roadmap phase through the sequencing of innovation projects to both satisfy customer demand and develop capabilities for future projects. As discussed in the last chapter, this is more complex than selecting shots in pool because "fog" almost always obscures everything, most particularly customer tastes and the competitors' pipelines of future projects. However, by modularizing the dependence of the firm's capabilities upon one another, the innovation executive can at least minimize the risk of a derailed roadmap arising from internal sources.

At the same time, expert pool players can continuously sink one ball after another, thus denying an opponent the ability to sink any of his own balls. Similarly, if an innovation executive can deliver products quickly enough, she can shape the expectations of the market place. That is, by playing modular shots, a firm's ability to shape and visualize the evolution of market place improves, making future planning simpler and more effective, hence speeding up the rate at which you can execute the SROM cycle. One could argue that Apple has accomplished this through the introduction of its iPod product family.

Chain of Interactions in a Portfolio: Buffering

How does one minimize the interactions between capabilities being developed on various ongoing projects? In maneuver warfare this question is akin to the need for ever-smaller standalone military units, such as a platoon working in a remote location, to be given sufficient resources and contingencies to enhance the chance that this unit can meet their objectives or at come close to it. Traditionally, the smallest standalone unit capable of independent operations was a division, which consists of approximately 15,000 soldiers. Nowadays, the U.S. Marines basic standalone unit is the Marine Expeditionary Unit (MEU), comprised around an infantry battalion (approximately one-ninth of a division) reinforced with fighter/attack aircraft, helicopters, tanks, artillery, and engineers. Making smaller units capable of independent action promotes a certain degree of duplication of resources and "overkill" in order to provide flexibility in case of emergent events. This flexibility, while expensive, enables superior planning at an aggregate (i.e., higher) level of planning because such overkill increases the chances for a mission's success, making the action of any one MEU less likely to fail and thus more predictably supporting the effort of other combat units. Hence, the combat landscape becomes much less chaotic than it would otherwise be and more amenable for generals and commanders to shape it in a desirable manner.

How can this lesson be analogized into maneuver-driven competition in the business arena? One way is to create some sort of buffer, e.g., in time or resources, to ensure that individual projects do not need to steal resources from another project. For example, managers can create time buffers between successive projects so that the likelihood that a project interferes with the timing of other projects—and thus throws the engineers off their roadmap—is relatively small, say 5% of the time. Eliyahu Goldratt discusses methods for doing this extensively in his book *The Critical Chain*.[11] He suggests an even more subtle option, which is to create buffers between activities, both within and across projects, which are using the same capability. Another way to create a buffer is to deliberately overstaff critical tasks or projects, both in a manner analogous to the MEUs discussed earlier. In both cases, of course, strict discipline must be instilled to avoid the perils of Parkinson's Law, "Work expands to fill the time available to complete the project."[12] If the project takes less time and effort than planned, one can "time-share" some of the engineers with less time-sensitive but important processes such as developing next-generation engineering tools and shared processes (see Chap. 7 for further discussion). On the other hand, if a critical task or project requires more effort than expected, it is far better to overstaff the project from the onset rather than try to add staff once a project or a critical project task is behind schedule. These last-minute actions often fall afoul of Brook's Law: "Adding staff to a late project makes it later."[13]

Another possible way is to set aside a contingency team and to deploy them specifically to deal with emergencies or disruptions in tasks that cannot be handled by the currently assigned personnel for that project.[14] This team needs to be familiar with all the active projects at a given point in time. This option, while not easy to implement, does offer a cost saving opportunity. Overstaffing each critical task on each project may prove prohibitive because many of the tasks will not run into trouble, thus wasting the overstaffed resources. However, if an innovation executive considers a number of critical tasks over all projects in the entire portfolio, she can "pool the risk" by more accurately projecting the fraction of tasks that will have problems. She can use this improved estimate to limit the number of extra staff allocated for problems across the entire portfolio of projects. The idea of such risk pooling of not new: an insurance agent in Milwaukee writing a policy does not knows in a given year whether, for example, any individual car may be destroyed in an accident, forcing an individual to buy a new car. Setting aside $15,000 for each insured care would

[11] Goldratt, E.: The Critical Chain. North River Press (1997).

[12] Northcote Parkinson published this law as a part of an essay in the Economist (1955). For an analysis of its implications, see:

• Gutierrez, G.J., Kouvelis, P.: Parkinson's law and its implications for project management. Manage. Sci. (1991).

[13] Brooks, F.: The Mythical Man-month. Anniversary ed. In 1995 by Addison-Wesley Longman, Boston (1957).

[14] This possibility was initially suggested to us by Professor Christoph Loch during the development of our work on hierarchical planning. We are grateful for this insight.

be prohibitive. This is, however, not an issue for an insurance company that supports the agent in Milwaukee because it is not planning for any single individual car, but rather for a pool of thousands of cars. Based on past history, the company knows that normally between 2% and 3% cars will be involved in accidents each year. The company can then set aside 3% of the value of all the cars insured and can be fairly confident that enough money has been set aside for normal road accident coverage. In a project context, this means that the overstaffing required to support a contingency team to ensure that 95% of all projects are completed on schedule will require many fewer employees than if each project is overstaffed individually. But, this contingency staff (also known as the bailout team or trouble shooters) must be composed of highly experienced generalists that can handle an array of problems.

Modularizing Capabilities

Another way to prevent many of these problems is to segregate risks in the roadmap across different projects so as to minimize the chances of any potential interaction between risks in a single project, because interacting risks often result in problems that are proved particularly hard to resolve. For example, some firms have a policy in place to avoid simultaneous changes on different generations of projects, or replication of changes across multiple locations. Intel is well known for introducing a new microprocessor product only on a proven manufacturing process. Only after the initial set of design "bugs" are worked out for the new microprocessor on the proven process, will the microprocessor be moved to a new process to improve production efficiency.[15] Intel's thinking behind this possibility is that, if a bug were to be found on a new microprocessor produced by a new process, the cause of the bug could prove to be either the new microprocessor or the new process, or perhaps some interaction between the two. This ambiguity makes it harder to track down and resolve the bug. By segregating the two risks, Intel can, on average, make each project more predictable and thus more amenable to planning. For much the same reason, World Motors has developed the new engine controller and the new engine on different generations of products. In the manner, segregating project risk facilitates faster execution of the SROM cycle by making the problems associated with each project easier to resolve and hence, ultimately, more predictable. This in turn will help prevent the emergence of an innovation butterfly that could be harmful to the firm. This brings up the difference between risk and uncertainty in sharp focus while making strategic product development choices. We illustrate this point by returning to playing pool analogy.

---

[15] For a discussion of Intel's Copy Exactly! Technology transfer method, see:

- Terwiesch, C., Xu, Y.: The copy-exactly ramp-up strategy: trading-off learning with process change. IEEE Trans. Eng. Manage. (2004).

As stated earlier, a key aspect of playing pool well is to limit the interaction of the various billiard balls on the playing surface. This is done most easily and reliably by modular play that prevents the interaction with specific billiard balls so as to avoid moving balls into an unfavorable position, i.e., to avoid disrupting their "roadmap." Similarly, modularization can be implemented in the management of the capabilities of the workforce through cross training, embedding capabilities in artifacts, by moving to shorter projects, or by implementing agile practices in conventional projects.

Cross-Training Personnel

Personnel can be cross trained in multiple capabilities so that they can be reallocated quickly during a project as disruptions occur (and they always do). There are a number of ways to do this cross training. Many firms train an individual engineer deeply in one particular technical skill. For example, at the real world firm that inspired the World Motors example, engine controller programming routines are learnt primarily from training provided by their CAD/programming vendors. These controller engineers may also get trained on in a number of related tasks such as engine design, circuit design, electronics packaging engineering, and general mechanical engineering. This is referred to as "T-type training" in which the cross of the "T" represents the individual's familiarization with a number of technical areas, and the central ascender represents an Engineer's "deep" expertise in one technical area.[16] Another method, espoused by Toyota, requires training each engineer deeply in two related tasks, such as electronics controller programming *and* circuit design or perhaps two dissimilar tasks, such as circuit design and the development and detailed design of an allied manufacturing process. This is referred to as "pi-type-training," because the Greek letter pi (Π) has two ascenders representing the two deep areas of an individual's expertise. Related to pi-type training is the fact that certain fields, such as circuit design and controller programming, are partially transferable. For example, a circuit designer can, in a pinch, do some programming.[17]

Another, and perhaps even more useful type of training for modularizing capabilities, is for innovation workers to all learn the same systems engineering standards and protocols used throughout the firm. This would enable all project engineers to communicate and interact with one another enabling a "plug-and-play"

---

[16] For a detailed discussion of the types of competence (including the T shape), see Dr. Daniel E. Whitney's online papers. For instance,

- Whitney, D.E.: CAD and product development in the US automobile industry, Available under the heading, "Auto Industry Perspective," at http://esd.mit.edu/esd_books/whitney/whitney_online.html.

[17] One of the authors came across the "T" versus "Pi" terminology during the course of his work in the automotive industry.

capability. However, the effort needed to achieve this shared exposure can be time consuming and in many settings, this adds significant burden to the engineering effort and escalates the cost. However, given the unpredictable nature of innovation butterflies and allied cost overruns associated with their effects, it may be possible to cost justify this overhead of shared know-how. To use a weather example, no matter how certain a weather forecast might be for good weather, no Boy or Girl Scout Troop goes on foregoes the added pound of a rain jacket in their backpack because weather predictions are just that, educated guesses, not certainties. Intangible benefits of such work design is that it keeps the engineers motivated in their domain and makes them more adaptive when they are faced with new learning opportunities.[18]

Modular Actions and Management of Risk

Ultimately, the goal of modular actions as described throughout this chapter is to enable speedier execution of the SROM cycle. By taking a modular action, such as cross-training personnel, an innovation firm increases the predictability of changes in the innovation landscape, which enables quicker scouting and less roadmap modification. A different way to think about this choice is that when an innovation butterfly arises, modular actions create options for rapidly recombining the various capabilities within a firm analogous to Baldwin and Clark (2000)'s modular operators, such as substitution, augmentation, or inversion between modular components.[19] Developing capabilities, of course, takes much time and often must be done in a sequence of steps. However, modular actions can be incorporated within the SROM cycle in order to plan a capability portfolio strategy that can better manage risk.

Perhaps the best way to enable this for each SROM cycle is to develop a number of plausible scenarios for the innovation system and then determine an alternate roadmap for each scenario along with the needed capabilities to execute that road-map (stylized examples of such scenarios and the evolution of roadmap are provided in the appendix). Then capabilities can be classified (just like in the commonly used product-inventory planning methodology) as either A, B, or C capabilities. "A" capabilities are those that are certain to be needed during the most likely scenarios. That is, it would be very difficult to create the road map without having these in-house capabilities. We view these to be "integral" to the firm's portfolio. Hence, these capabilities must either exist in-house or be developed sufficiently prior to the time of need. Other capabilities that are either less likely to be needed or less

---

[18] Glen, P., Maister, D.H., Bennis, W.G.: Leading Geeks: How to Manage and Lead the People Who Deliver Technology. Jossey-Bass (2002).

[19] Clark, K.B., Baldwin, C.Y.: have defined a set of operators. For instance, "splitting shrinks the "footprint" of each task or process," and "augmentation" introduces a new module that plugs into existing interface (2000).

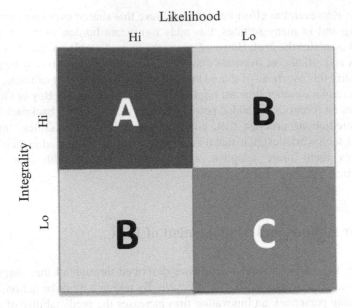

Fig. 6.4 Integrality and likelihood of capability needs

integral to projected future project portfolios can be classified as "B" capabilities. The rest of the capabilities are graded as "C" category, which is assumed to be readily available for acquisition from the market place on an as-needed basis.

With "B" capabilities, the firm develops an understanding of the capability and perhaps even the capacity to develop and deploy that capability in a small number of projects. However, the bulk of project tasks requiring "B" capability will be provided by outside firms. The development of a "B" capability then essentially becomes a hedge against the unpredictability in the innovation system arising from its complex nature. By having a minimal capacity for "B" capabilities in-house, the firm can integrate other firms' efforts through partnering or through outsourcing effectively within its own product portfolio. Furthermore, if the "B" capability in question evolves over time to become essential to the firm's roadmap or other likely scenarios, then this "B" capability could eventually serve as the starting point toward developing a full-blown "A" capability. A matrix displaying the relationship of a capabilities likelihood, integrality, and classification is presented in Fig. 6.4.

A more complete example of the data analytics associated with SROM cycle centering on road mapping and capability planning is presented in Appendix B.

Embedding Contingent Capabilities in Artifacts

We have argued that people are the carriers of know-how and capabilities in a firm and that managing the interactions among requirements for people allows a firm to track and minimize the unwanted effects of butterflies. There exists an option for protecting against butterflies embedding these capabilities into artifacts rather than personnel.[20] An example of such an artifact is the tools, whether they may be actual software, reference manuals, or "wikis," that allow a firm's knowledge to be accessed on an as-needed basis by engineers of any field or capability. This proves to be a more difficult process than it first appears because of the "tribal" or "tacit" knowledge that technical personnel possess, but, for various reasons, rarely write down. For example, automotive engineers generally know that they must always filter or shield out radio-wave interference emanating from the spark plugs in auto-motive electronics products and that the "ground" on a vehicle can vary by up to 1 Volt in different parts of the vehicle. Yet this information is not commonly written out in manuals. It is merely passed from teacher to student or mentor to apprentice during training. The challenges and benefits of such information, thus capabilities within tools and shared processes rather than personnel are further explored in the next chapter.

---

[20] Carlile, P.: A pragmatic view of knowledge and boundaries: boundary objects in new product development. Organ. Sci. **13**(4) (2002).

Chapter 7
Plug-and-Play Capabilities

Secure Technologies, Staff, and Processes on Demand to Support Innovation

This is a world of processes, not a world of things.

Margaret J. Wheatley[1]

As we described earlier, the SROM cycle can be used to shape the customer needs and to exploit the innovation butterflies, thus setting the scene for successful maneuver-driven competition. In the last chapter, we described how modularizing capability risk across the innovation portfolio enables a faster tempo for the SROM cycle, which increases the chances that an innovation firm can shape the innovation system to its own benefit. We now describe how common processes and tools, such as talent management, workforce planning, shared business processes, and information management systems can also help accomplish these ambitions. We also note that implementing such "processes" can turn out to be a double-edged sword—some view this to be a straightjacket that crimps the creative process.

Returning to the realm of maneuver-warfare, successful military organizations drive ceaselessly toward the development of shared doctrine and processes. The reasons for this are threefold: one is to ensure—to the extent that anyone can do so in such an uncertain environment—that the processes of a military organization, including all of its units, are consistent. The more that the soldiers or marines are trained to perform these processes consistently, the more likely they can respond with them effectively under duress, which increases their chances to carry out a mission successfully. Second, process consistency enables senior commanding officers and sister units to better guess what a unit will do under a particular set of circumstances. This simplifies planning by reducing the number of scenarios that must be planned for, thus speeding up the SROM cycle. Finally, if a unit, or parts of it, needs to be combined with another unit, soldiers can work together much more quickly

---

[1] Wheatley, M.J.: Leadership and the New Science: Order in a Chaotic World. Barrett-Koehler, San Francisco, p. 68 (1994).

E.G. Anderson and N.R. Joglekar, *The Innovation Butterfly*,
Understanding Complex Systems, DOI 10.1007/978-1-4614-3131-2_7,
© NECSI Cambridge/Massachusetts 2012

and effectively than if each unit did things in different ways. To some extent then, the standardization of processes is an extension of the modularization of capabilities described in the previous chapter.

In the context of innovation, a *shared process* is a sequence of standardized tasks that are executed across several processes in a distributed product development project. Sometimes these processes can be temporally grouped together, in which case they are oftentimes executed by a group of specialist individuals, such as bench scientists in a biotech industry or mechanical engineers in an automotive design group. Whether such shared processes can be set up is influenced by organizational tensions and the principle of distributed providence, because the structuring of such processes not only affects task decomposition and integration but also may potentially result in job reduction or the shut-down of R&D locations. Over the long haul these processes become part of organizational culture—in essence, they become the manner in which much of the work of innovation projects gets done.[2]

We begin this section with a case study from the medical devices industry and show that the capability planning and oversight processes operate under many uncertainties. We then review best-in-class tools and technologies that allow managers to leverage such shared processes to facilitate planning in the face of emergent uncertainties.

Case Study: Staffing and Shared Processes at MedDev[3]

This case study focuses on a large healthcare company, MedDev, which develops, produces, and markets products for disease monitoring and control. These medical products have evolved over time through a sequence of product-line enhancements followed by the occasional introduction of new product families. Historically, MedDev has derived its revenues from sales of its devices to individuals and hospitals. But, its marketing specialists convinced management that a shift toward

---

[2]The evolution of processes and staffing in innovation settings cannot studied without explicit attention to knowledge creation and management issues. We have elected not to address these issues explicitly for brevity. For formal studies in this domain, see:

- Anderson, E.G.: Managing the impact of high market growth and learning on knowledge worker productivity and service quality. Eur. J. Oper. Res. **134**(3), 508–524 (2001).
- Carrillo, J., Gaimon, C.: Managing knowledge-based resource capabilities under uncertainty. Manage. Sci. **50**(11):1504–1518 (2004).
- Ozkan, G., Gaimon, C., Kavadias, S.: Knowledge Management Strategies for Product and Process Design Teams. ssrn.com/abstract=1520771 (2009).

[3] This is an abridged version of the case with one new exhibit. For allied discussion of the underlying case, see:

- Joglekar, N.R., Rosenthal, S.R.: Coordination of design supply chains for bundling physical and software products. J. Prod. Innovat. Manage. **20**, 374–390 (2003).

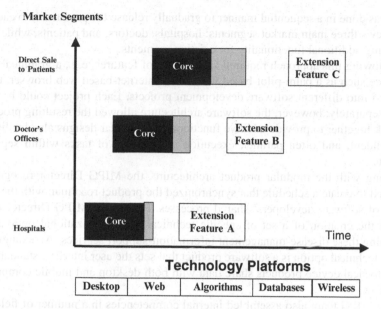

Fig. 7.1 Modular product line growth

providing value-added software features bundled with its devices was necessary for continued growth and revenue streams.

Anticipating that such a shift in product strategy would call for new organizational and management approaches, MedDev's management decided to assign this innovation responsibility to a new development group assembled solely for this purpose. The rationale for setting up a separate group was that this way MedDev could create an entrepreneurial focus in one area—software design and development—in which it had lacked extensive expertise. The group was named the Medical Information Product Group (MIPG). The Vice President of the R&D function also created a new position, Director of MIPG, whose first task was to identify the key operations that the software would be expected to perform. Traditionally, R&D groups at MedDev constituted of bench scientists with expertise in areas such biochemistry. The MIPG product development process would require the addition of new competencies, including R&D staff with expertise in software architecture, integration of software with medical solutions, and clinical testing and marketing of the software offering. To facilitate this, the MIPG Director decided to follow a modular approach to control the design, development, and testing of a sequence of new software development projects as shown in Fig. 7.1. With modular development, each project can be carried out in standalone manner, and then the outcomes are integrated over time as new modules come on line, much like a LEGO® set, in which additional parts can be bought and integrated over time. Initially, the core hardware was to be introduced at hospitals as shown on lower left side of Fig. 7.1. Over time, the core was expanded (as shown in gray) by the addition the application programming interfaces (APIs), which contained extension features A, B, and C.

This was done in a sequential manner to gradually release the products into each of MedDev's three main market segments: hospitals, doctors, and patients, while also providing additional functionality for existing segments.

Following this approach required various sets of features, e.g., a graphical user interface such as a palm-pilot-based view or an internet-based web browser, to be grouped into different software development projects. Each project could be executed separately; however, the software architecture allowed the resulting products to work together to provide additive functionality. Modular designs also facilitated independent, and often parallel, execution and control of tasks within separate projects.

Along with the modular product architecture, the MIPG Director recognized the need to create a schedule that synchronized the product road map with the evolution of software developers' shared processes. Hence, the MIPG Director envisioned the creation of a set of specific technical and organizational options for designing these disease-management information-support services. An example of such a technical option is a software product that sets the user interface standard for how physical device products share data with both desktop and mobile computing platforms.

The MIPG team also assembled internal competencies in a number of fields of specialization to effectively manage the interface with its development partners and suppliers. These resulting organizational processes involved establishing software architectures, customer requirements and parts specifications, external evaluations (e.g., the clinical aspect of the software usage), failure mode effects analysis (FMEA), human factor testing, design of graphical user interfaces, and validation protocols. Each process had associated technical and administrative tasks including problem solving and specification writing. The administrative tasks included project management, design reviews, market research coordination, contract negotiations, and performance reviews.

Setting up and managing these processes also required another set of competencies: maintaining a regular update on the inventory of competencies as scientists got on and off projects, and when possible, assigning certain individuals to perform similar tasks on multiple projects and processes as shown in Fig. 7.2 (diagram format, cell titles, and individual names altered to preserve privacy).

The roadmap in Fig. 7.1 was a document that was shared by all senior managers, and simplified version of the roadmap was also shown to the customers. Figure 7.2 was maintained by the director of MIPG using a modified version of excel spreadsheet for her own usage. It captured the cells that seem to face high risks (by identifying them with three stars "***"). Another version of this diagram also contained the percentage of time that an individual was assigned to a project, along with their progress status (i.e., percent complete and projected completion date) on each cell. It also identified some shared resources (with the symbol §) because the person's headcount resided outside the group, but the person shared his or her time with MIPG. The director of MIPG updated this spreadsheet weekly and conducted simple analytics to keep track of the overall pattern of completion, dependencies and delays, in order to identify emergent trends. The director also kept track of capability of the individuals who worked on various steps in these processes.

		Products (partial list)			
		Patient's Handheld	Patient's Desktop	Dr's Office	Hospital
Processes (partial list)	Cable Interface	Geri + Intern	Geri, Don	Don**	NA
	User Interface	Outsourced to partner # 1 with Tom as Overseer ***	Tom	Jill	Outsourced to partner # 2 with Jill as Overseer **
	Database Design	Anil	Jie	Anil	Jie
	Platform Linkage	Vanessa	Vanessa	Vanessa**	Vanessa + Teri §
	Hub /Network Standards	Avi §	Avi §	Avi §	Avi §
	Web Interface	Outsourced to partner # 1	Outsourced to partner # 1	Outsourced to partner # 2	Sign contract with partner # 2
	FMEA	Duncan***	Duncan***	Not Staffed**	Not needed yet
	Clinical Testing	CT Team A	CT Team B	CT Team C	CT Team C

***High Risk   **Medium Risk   §Shared Resource

Fig. 7.2 Mapping resources across products and processes

The MIPG Director noted that the collection of these types of maps was incredibly useful to her during hiring, developing and even pruning the capabilities for the ever-shifting needs of the MIPG team. For instance, the MIPG team had to work across variety of inter- and intraorganizational interfaces. As the team moved from product extension A to B to C, and addressed new market segments, development of API interfaces required ongoing collaboration with external software developers; links with external software-standard-setting bodies such as the Connectivity Industry Consortium (CIC); interactions with prospective software users, doctors and hospitals; plus interaction with MedDev's physical product R&D teams. A number of potential butterflies were created during the definition of module content and specification of boundaries, which required adjustments to the roadmap, redefinition or adaption of tasks within processes, and reassignment of specialists to these processes. Accordingly, the MIPG's evolving agenda had to include building new boundary-spanning skills for the team to coordinate across organizations. Within the MedDev context, boundary-spanning tasks required "scouting" for innovation patterns and market demand, negotiating and awarding contracts, developing and communicating specifications and program status, learning the suppliers' technical competencies and organizational processes, understanding and controlling uneven cost structures across tasks and suppliers, and managing rework and queuing delays across organizational boundaries.

Dealing with these issues required hiring personnel with generic project management skills and then encouraging them to develop new expertise either through hands-on learning or through specialized training, including how to negotiate contracts and master the intricacies of various stage-gate processes put in place by their development partners. While explaining why she hired a certain project manager, the leader of MIPG commented: "He was trained in the military and is good at being the glue." She also suggested that "the more 'languages' one can speak the better it is," later explaining that the term language in this context referred to the project managers' facility in communicating the technical and administrative languages of the various project members and their specialist disciplines.

These observations are not unique to the MedDev business context. We have observed similar types of integration and coordination challenges when the number of interruptions, i.e., events that need special handling, rises with the degree of distributed work either geographically, or across firm boundaries; this can often lead to a vicious cycle of interruptions.[4] Selection, nurturing, and continual assignment of key staff, and enhancing their ability and willingness to share their capabilities across active product and process lines then become a high leverage activities for the planners, who are charged with the organization of distributed innovation processes. A common technique used toward to facilitate this activity is aggregate workforce planning, which we describe next.

Aggregate Workforce Planning[5]

Most R&D organizations carry out aggregate workforce planning on an annual basis, typically in conjunction with the annual budget planning process. The goal of such processes is to acquire and allocate each of the resources needed to execute a project, primarily skilled technical employees such as engineers and programmers.

---

[4] For a discussion of integration issues, see:

- Parker, G.G., Anderson, E.G.: From buyer to integrator: the transformation of the supply chain manager in the vertically disintegrating firm. Prod. Oper. Manage. 11(1), 75–91 (2002).

For a discussion of interruptions during distributed development, see:

- Anderson, E., Davis-Blake, A., Erzurumlu, S., Joglekar, N., Parker, G.: Managing Distributed Product Development across Organizational Boundaries, Chapter 10. In: Loch, C., Kavadias, S. (eds.) The Handbook of New Product Development Management, Butterworth–Heineman, Oxford (2008).

[5] For foundational, and model-based, discussion of aggregate planning in production setting, see: Holt, Modigliani, Muth, and Simon (1960). For companion piece that reviews the literature and lays out the challenges in implementing these ideas in innovation domain see Anderson and Joglekar (2005).

- Holt, C. C., Modigliani, F., Muth, J.F., Simon, H.A.: Planning production, inventories, and work force. Prentice-Hall, Englewood Cliffs, NJ: (1960).
- Anderson, E.G., Joglekar, N.R.: A framework for hierarchical product development planning. Prod. Oper. Manage. 14(3), 344–361 (2005).

Hiring, training, and deploying these employees generally involve significant costs and time delays. Another decision for firms with various design centers located in several locations (e.g., the USA, Europe, and India) is where to base a certain project. Such decisions generally have a significant impact on cost, timing, and perhaps quality.

If resources for a particular location are expensive because of high wage rates or if there are any budget shortfalls, the required development capacity may be obtained through outsourcing to a third-party development firm, though this may result in higher coordination costs and/or lengthened completion time or lower product quality. The time horizons and lags involved in these decisions may be shorter or longer than those in portfolio selection depending on the industry. During normal periods (those not involving financial duress, downsizing, or hiring freezes), these decisions generally will be made at a middle-management level. Indeed, because the aggregate number of employees is fixed and many employees work on multiple projects, the aggregate planning process affects all the projects within an organization.

We now highlight two aspects of the aggregate planning. The first is the need to align this process with the SROM cycle, which we have introduced in Chaps. 5 and 6, and the second is the fact that data obtained during the aggregation process is far from perfectly accurate. This contributes to the "fog" in the innovation system described earlier, which can slow down the SROM cycle because managers do not want to assign employees to projects until it is fairly certain that they will not need to be moved. The reason for this is that frequent personnel reassignments often result in lowering of morale and productivity.

In many traditional R&D organizations, aggregate workforce planning is delegated to human resource (HR specialists) and is considered a staffing function. This might be reasonable in planning for a relatively stable setting, such as a large administrative team charged with processing taxes or loan applications. However, this approach is not useful in a product or service innovation settings because the nature of innovation tasks overwhelm the organization with potential innovation butterflies from a variety of sources: markets, government regulations, processes, creative efforts, and behavioral biases as we showed in the market-capability loops. Additionally, as discussed earlier in Chap. 2, innovative products are the result of innovative capabilities (and innovative combinations of various capabilities). Hence, much of the roadmap and orchestration phases of the SROM cycle require rapid interaction at the aggregate product and staffing planning level, and the staffing functions (especially task assignments) cannot be delegated to an HR specialist who may not have the relevant information or be capable of understanding how aggregate product plans are evolving. For instance, in the MedDev case, the Director of MIPG took on the responsibility for aggregate product planning as well as the aggregate workforce planning.

We note that MedDev is by no means alone in linking product line and business planning with workforce planning. Even a very large organization such as IBM Global Services has gone to great lengths to develop and maintain inventories of its competencies and to introduce highly sophisticated software environments that create marketplaces for jobs (using a software tool called GOM), for staffing projects (using another tool called professional market place), and for contracting outside labor (using a tool called CITRUS).[6] These tools gather data, and these data can be exchanged, aggregated, and communicated through appropriate interfaces. Further, trend analyses can be conducted to provide valuable forecasts on the gaps between talent demand and supply and their linkage into the business needs for an innovative organization. Allied analytics capabilities are further discussed in the appendix.

A related issue is the quality of data in such settings. In our view, data within shared processes, particularly for aggregate planning, contains much measurement error, which contributes to the general fog surrounding innovation plan. Measurement error in particular is often referred to as "noise" because it is similar conceptually to the static or "white noise" one can often hear in when one is trying to listen to a radio. Such "white noise" can be so loud that the listener can no longer hear what a radio announcer is saying. (This can be very frustrating during football games!) A majority of research on aggregate planning assumes that there is no "noise" with respect to measurements of the progress being made on various projects as they are executed; in essence, it is assumed that the information on progress status is perfect, lacking any error resulting from such sources as underestimating the time required to complete current tasks or collect information through information systems. However, in a set of projects that we have observed, the mean error in measurements of project progress status data (when compared with an audited sample) is about 28%.[7] Availability of progress data uncontaminated by noise is generally assumed to be a critical planning requirement—best practices in project and portfolio call for the use of up-to-date progress status data to make decisions such as project resource allocation.[8] Even when innovation tasks are well specified, however, it is very difficult to establish the progress status with certainty in terms of hours of work remaining. In some instances, the uncertainty in the progress is caused by variations in productivity, such as holidays, absences, and by changes in end-customer demands impacting the work required. In other cases, uncertainty in progress status is caused by information hiding and the modular architecture of the development process.[9] Other issues that compound the difficulty of tracking project status are errors inherent in information systems when data are aggregated across many projects—e.g., the resource needs for small, but critical, task may be hidden when the dataset includes much larger tasks, with less pressing needs, because of the scaling effects. Finally, owing to the principle of distributed providence, some participants may willfully misrepresent their progress status.[10]

---

[6] Young, M.B.: Strategic Workforce Planning in Global Organizations, Research Report. The Conference Board (2010).

[7] Shankaranarayan, G., Joglekar, N.R., Anderson, E.G.: Managing Accuracy of Project Data in a Distributed Project Setting. Proceedings of ICIQ, Little Rock, AR (2010).

[8] Verzuh, E.: The Fast Forward MBA in Project Management, 2nd edn. Wiley, New York (2005).

[9] Yassine, A., Joglekar, N., Braha, D., Eppinger, S., Whitney, D.: Information hiding in product development: the design churn effect. Res. Eng. Des. **14**, 145–161 (2003).

[10] Ford, D, Sterman, J.: The Liar's club: concealing rework in concurrent development. Concurr. Eng. Res. Appl. **11**(3):211–219 (2003).

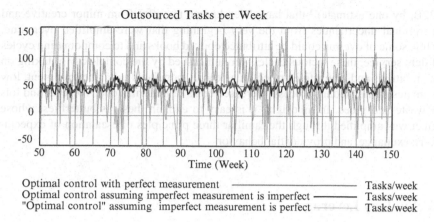

Optimal control with perfect measurement ——————————— Tasks/week
Optimal control assuming imperfect measurement is imperfect ——— Tasks/week
"Optimal control" assuming imperfect measurement is perfect ——— Tasks/week

Fig. 7.3 Effect of filtering noise from the data used for aggregate planning

Such noise in the status of project completion is especially problematic because senior R&D managers in many firms rely on such data for oversight and adjustments of their portfolio.[11] If they are lagging behind, and do not have capacity, they assign tasks to their outsourcing partners. They also rely on these data to indicate availability of new products to their customers. Portfolio oversight is particularly relevant in professional services settings, such as management and IT consulting firms, because their staffing plans are based on the underlying evolution of tasks (or projects).

In a model-based study, we calculated that measurement noise at the level we observed can easily add double-digit-percentage penalties to staffing costs.[12] Needless to say, the presence of innovation butterflies can easily amplify the problems created by noise and increase costs by up to an order of magnitude. Figure 7.3 provides graphical evidence of such amplification. It shows that without measurement noise the management team would have outsourced about 50 tasks/week to keep their portfolio in balance.

Under otherwise identical test conditions, the managers will have to go out at seek outside contractors up to 150 tasks/week in the presence of noise. While the impact of these claims from the model may seem huge, we draw the reader's attention to the escalation of budgets in mega projects such as Boston's Big Dig. The project involved construction of a major transportation artery in downtown Boston, along with bridges and infrastructure improvement. The project started out with an initial estimate at $6B (that ballooned to over $14B and may actually be a staggering

---

[11] Wheelwright, S., Clark, K.: Revolutionizing Product Development: Quantum (1992).
Leaps in Speed, Efficiency, and Quality. The Free Press, New York.

[12] Shankaranarayan, G., Joglekar, N.R., Anderson, E.G.: Managing Accuracy of Project Data in a Distributed Project Setting. Proceedings of ICIQ, Little Rock, AR (2010)

$22B, by one estimate)[13] that have been known to suffer from minor creative and operational uncertainties from the very beginning that were amplified over time. While some of overrun could be attributed cooked books and to compounding cycles of debt service, these vicious cycles were unleashed by the complexity of taking on an ambitious and innovative project in a densely populated urban environment, low estimates in the bidding process, increased costs of materials over 10 years and lots of waste and greed. In hindsight, it is easy to explain these as butterflies, whose effect was amplified through the familiar three principles of escalation of expectations, exchange, and providential behavior.

Technology Leverage

To the extent that such "noise" can be removed from the system, by using methodologies such as Total Data Quality Management, the SROM cycle can be sped up and the innovation firm made more agile.[14] Improving data quality is not the only technical opportunity. Processes for carrying out innovative tasks also evolve over through R&D effort. Historically, enhancements to the technical aspects of product development processes have been captured in software environments such as computer-aided design (CAD) tools.

For instance, the Boeing 777 was the first commercial aircraft designed entirely on computer. Everything was created on a 3D CAD software system known as CATIA. This allowed a virtual 777 passenger aircraft to be assembled in simulation and enabled the process of checking for interferences and verifying proper fit of the thousands of parts before costly physical prototypes were manufactured. Boeing was initially skeptical of the capabilities of the CATIA system and built a mock-up of the aircraft's nose section to test the results. It was so successful that all further mock-ups were cancelled.[15] While the scales of technical complexity may differ, the leverage of technology is itself a necessary innovation to enable efficient processes settings ranging from the development of microprocessors to the processing of graphics and animation in the video game industry.

---

[13] For the impact of Boston's Big Dig project on rework and delays, see:

- Review Begins After Big Dig Tunnel Collapse. CNN.com. 2006-07-12. http://edition.cnn.com/2006/US/07/12/bigdigdeath.ap/index.html. Accessed 25 Jul 25.
- Big Dig's red ink engulfs state, Boston Globe, July 17, 2008. This article indicates that, in all, the project will cost an additional $7 billion in interest, bringing the total to a staggering $22 billion, according to a Globe review of hundreds of pages of state documents. It will not be paid off until 2038.

[14] See, for instance: http://web.mit.edu/tdqm/www/about.shtml.

[15] From aircraft.wikia.com/wiki/Boeing_777.

The advent of ubiquitous and easy-to-use web-based interfaces has hastened the trend toward IT-based collaboration across traditional organizational boundaries in a variety of industries. Such collaboration touches all facets of distributed development: problem solving required both as the component and at the system level. Other information technology-based refinements in this space include virtual customer interfaces, communication of development intent, synchronization of product lifecycle management tools, and exchange of bills of materials across organizational boundaries. In many industries, the use of IT tools is much more widespread for technical problem solving than for coordinating distributed development efforts. That the use of automated collaboration tools results in higher quality products resembles the paradox of information system productivity because personnel productivity gains are offset by a firm's tendency to deliver higher quality goods, *which require a more expensive manufacturing process.*[16]

Our colleague Professor Stefan Thomke has argued that productivity improvements with new development tools must be explicitly managed because they face pitfalls such as (1) their utilization as mere substitutes for existing practices and (2) the fact that they introduce additional interfaces into the innovation process.[17] Recent research in this area separates the effects of problem solving and administrative productivities and shows that each of these areas contains opportunities for managing complexity through the use of technology. The net effect of such tools upon the SROM cycle has been twofold. One is that in some industries, such technologies can create virtual prototypes for more effective project reviews by customers. The output of these reviews can then be used to adjust the innovation project at relatively low cost, thus improving the agile project management process. This is a relatively common practice in some software project settings. Hardware development projects are also instituting similar agile reviews. More frequent individual project reviews increase the rate at which the maneuver portion of the SROM portfolio cycle can be executed.

---

[16] In many industries the use of IS tools is much more widespread for technical problem solving than for coordinating distributed efforts. The use of automated collaboration tools results in a higher quality product (Joglekar and Whitney 1999) analogous to the information system productivity paradox (Brynjolfson and Hitt 1998): productivity gains are offset by a firm's tendency to deliver higher quality goods. Allied ideas are explored by Bardhan et al (2007) and Nambisan (2009).

- Brynjolfsson, E., Hitt, L. Paradox Lost? Firm-level evidence on the returns to information systems spending. In: Willcocks, L., Lester, S. (eds.) Beyond the IT Productivity Paradox: Assessment Issues. McGraw Hill, Maidenhead (1998). Reprinted from Management Science, 1996.
- Joglekar, N.R., Whitney, D.E.: Automation Usage Pattern during Complex Electro Mechanical Product Development. MIT Center for Technology Policy and Industrial Development Report, prepared under contract for the US Air Force Research Laboratory (US-AFRL) (1999).
- Bardhan, I.R., Krishnan, V., Lin, S.: Project performance and the enabling role of information technology: an exploratory study on the role of alignment. Manuf. Serv. Oper. Manage **9**(14), 579–595 (2007).
- Nambisan, S.: Information Technology and Product Development. Springer (2009).

[17] Thomke, S.: Capturing the real value of innovation tools. MIT Sloan Manage. Rev. **47**(2), (2006).

Despite these advances in measuring and handling technology leverage, they still remain very much an open arena of inquiry, particularly with respect to the role of innovation leadership in guiding the development of such processes.

To recap, in this section, we have presented several tools for managing the complexity of the innovation system. In particular, we described the process of maneuver-driven competition and several tools required to support it: using information scaling to shift to the portfolio level of management, employing the Scout–Roadmap–Orchestrate–Maneuver (SROM) cycle, modularizing capability risk, and implementing plug-and-play capabilities. In the next section, we explore other leadership issues in the innovation system, particularly with respect to how innovation leaders can mold the culture of their firms by encouraging creativity as a tool to manage the complexity of the innovation system.

Part III
Agile and Distributed Leadership

The most dangerous phrase in the language is, 'We've always done it this way.'

Admiral Grace Hopper, USN[1]

We begin this section by pointing out that the system of maneuver warfare described in the previous section involves multiple leaders acting with some degree of autonomy at different levels of decision making: a general making aggregate strategic decisions at the staff level; a captain leading a company; or a sergeant making choices for his or her platoon in the field. Similarly, innovation leaders have special roles at multiple levels in business organizations: firm strategy and portfolio planning; oversight of an individual project or team; or execution of problem-solving challenges within a laboratory or design subteam. The ideas of innovation leadership that we present in this section embrace all of these levels. Anticipating and managing the effects of innovation butterflies by deft maneuvering through emergent innovation challenges, especially distributed innovation challenges, needs such leadership at multiple levels within an innovation organization and across all the links to its partners. Agile and distributed leadership only becomes more pressing as innovation chains become more distributed, with supplier organizations widely spread through different geographies, industries, and cultures.

In the *Fifth Discipline*, Peter Senge suggested that the innovation leader's new work is more akin to being an architect of a ship than its captain.[2] Senge's metaphor was novel at the time and remains enlightening in many types of innovation domains. Because this book focuses on innovation management in decentralized innovation settings, we can be a bit more specific in suggesting ways a leader of such development efforts can contribute to and steer innovation processes toward successful

---

[1] womenshistory.about.com/od/quotes/a/grace_hopper.htm

[2] Senge, P.: The Fifth Discipline: The Art & Practice of the Learning Organization. Doubleday Business (1994).

outcomes in a complex world buffeted by the randomness inherent in technology, the markets, performance, and human behavior.

In particular, we believe that the decentralization required by the complexity of the innovation system requires that leaders master elements of the skills of *both the architect* and *the ship's captain*. (We have explored both these roles to some extent in the discussion of capability road mapping and orchestration in Chaps. 5–7.) In addition, we believe that the innovation leader must act not only as an architect *and* a ship's captain within the innovation firm but also as something like the coach of a high-performing athletic team. This analogy will be particularly helpful when we begin to delve into the realm of creating a uniform culture as a foundation for the successful practice of maneuver-driven competition, particularly because most of us have had exposure to organized sports either as a player or as a fan (far more experience than most of us have had in warfare, or for that matter, with nautical captainship). Additionally, sports have an advantage when discussing business leadership because the effectiveness of sports teams is clearly measurable in every game and every season thanks to scores and standing, which strongly parallel quarterly reports in the world of business.

Most importantly, however, team culture is as important (perhaps, even more so) for the innovation firm as it is for athletic teams. As discussed earlier, decentralization of decision making, as necessary among project managers as among football or soccer players, leads to many negative behavioral effects. Some of these negative effects are in some sense rational, such as an individual subordinating the good of the overall organization to that of himself. Other aspects of negative behavior come from economically irrational, but no less powerful, urges such as ego building and striving to become a "rock star." Building a culture that is based on a shared "love of problem solving," cooperation, and "rewarding solid work with more challenging and creative work," rather than focusing on monetary incentives or promotions can remedy or at least ameliorate many of these issues. Thus, the leader's ability to establish clarity of roles and building a culture of individual accountability becomes crucially important in innovation an organization.

Similar analogies can also be seen in high performing athletic teams. Both these settings require certain sacrifices for the good of the organization on the part of individual problem solvers (e.g., scientist, engineers, or football players). Fortunately, many engineers (or players) have deep understanding of how to solve problems or implement maneuvers. They develop long memories of how problematic situations have been dealt with in the past.

Given this history, if one sort of capability becomes nonessential, it may be in the firm's (or a team's) best interests to try to retain the people who made up that capability by converting them into another role, even if this takes some time and effort. This is not to say that the firm must protect everyone, only those who have made a real commitment to the culture. This is especially true for the middle management and technical "gray-hairs"—the innovation equivalents of grizzled veterans on professional sports teams—within the organization to acquire a deep understanding of how the firm (or team) works and the informal social network that supports it. Not only are they the people who are the carriers of the institutional memory of the successes and failures, but they are also in a position to shape the culture of an organization at least as strongly as an innovation executive, for example, by becoming "enforcers" who turn the ethos of good behavior (or its opposite) into a self-perpetuating reality. Similar behavior has been exploited by good head coaches in all sports since time immemorial to minimize the risks arising from poor individual behavior. In Chap. 10, we use the example of the head coach of an (American) football team as an analogy to argue that shaping the culture of a company through its norms, values, and procedures may be the most powerful "management" lever of all. Hence, we expand on Senge's metaphor within the realm of decentralized innovation systems to argue that a leader of an outstanding innovation firm must act not only as its *architect*, but also its *captain*, and moreover, its *coach*.

While discussing architects, we draw for illustrative purposes upon the careers of American architect and visionary Frank Lloyd Wright (1867–1959) and famous British naval architect Isambard Kingdom Brunel (1806–1859). The discussion is not limited either to buildings, or ship design, but extends into the broader field of systems architecting.[3]

For inspiration with respect to captaining a ship on a voyage of exploration (which has many difficulties analogous to those faced by innovation leaders), we use the career of Captain James Cook (1728–1779), the famous British explorer, upon whose exploits the character of James Kirk in *Star Trek* was based. In an astounding career that made him a fellow of the Royal Society, Captain Cook circumnavigated Antarctica, charted the east coast of Australia, and was the first European to "discover" the Hawaiian Islands. Interestingly, compared with previous explorers, Captain Cook painstakingly prepared for his expeditions and worked to enhance the survival rate of his crew both materially and by shaping his crew's culture. (In fact, he was arguably the first explorer of importance to bring a majority of his crew back home safely.) In particular, while he was known as being relatively lenient on discipline, he was quite strict on enforcing his crew's cleanliness and a diet laced with scurvy-preventatives such as sauerkraut. He was also willing to test and adopt new technology such as the chronometer method of measuring latitude that would simplify and improve his navigation measurements, which in voyages in uncharted waters could literally mean the difference between life and death.

Finally, in the realm of sports, we will draw upon the careers of two of the most successful American National Football League head coaches of all time: Bill Walsh of the San Francisco 49ers and Bill Belichick of the New England Patriots. While the pair had slightly differing views on what constituted an effective culture, they both had strong views that culture was important to winning games and employed "grayhairs" to help develop and maintain it. Moreover, they effectively integrated the strong cultures they developed with the other aspects of coaching, such as architecting the team, designing specific game strategies for specific opponents, and making adaptive adjustments on the field during game.

---

[3]For a discussion of systems architecture, see:

Eberhardt Rechtin, E.: Systems Architecting: Creating and Building Complex Systems. Prentice Hall, Englewood Cliffs, NJ (1999).

Chapter 8
The Leader as an Architect

The Architect Must Be a Prophet

> The architect must be a prophet ... a prophet in the true sense of
> the term ... if he can't see at least ten years ahead, don't call
> him an architect.
>
> Frank Lloyd Wright[1]

As stated in the introduction to this section, Peter Senge in the *Fifth Discipline* asked organization leaders to consider that the most "influential" position on a ship might not be the captain, but rather the ship's architect. In this chapter, we draw upon examples from a variety of settings, ranging from the history of ship design, revolutionary ideas tried out in the built environment (buildings, urban design, etc.) by Frank Lloyd Wright, to more recent and agile instantiations of architecture in the software industry. In doing so, we argue that to an organization engaged in distributed innovation needs leadership at multiple levels within many organizations. To an extent, every innovation leader in such settings inherently plays the role of an architect by exploiting emerging scientific capabilities, by having the vision for change, and by purposefully decomposing and integrating the interactions among key human elements on behalf of developers and customers during the innovation process.

Exploiting Emergent Scientific Capabilities

Larrie Ferreiro, a naval architect and historian, has argued that the history of modern ship design is tightly linked with the scientific revolution during a period when Great Britain became a maritime and industrial power, and its engineers used newfound scientific knowledge to help them solve the practical problems.[2] Such

---

[1] Wright quotation is from http://www.unitytemple-utrf.org/philosophy.html.

[2] Ferreiro, L.D.: Ships and Science: The Birth of Naval Architecture in the Scientific Revolution, 1600–1800. MIT Press, Cambridge, MA. (2010).

E.G. Anderson and N.R. Joglekar, *The Innovation Butterfly*,
Understanding Complex Systems, DOI 10.1007/978-1-4614-3131-2_8,
© NECSI Cambridge/Massachusetts 2012

studies provide elaborate evidence on the role of architecture in shaping the evolution of market wants and product performance described in Part I. For instance, some of the key innovations in the ship design came after Rudolf Diesel invented the diesel engine in 1892, and Sir Charles Parsons launched the "Turbinia" and demonstrated the potential of the steam turbine in 1894. As the maritime industries moved toward the adoption of oil as a fuel in place of coal, regulatory bodies and independent assurance institutions, such as Lloyd's Register, introduced rules for the burning and carriage of liquid fuel, for the adoption alternative propulsion mechanisms.[3] The innovation frontier in propulsion also opened up opportunities for improving the structural integrity of these ships through the introduction of steel instead of wrought iron. In order to further enhance this integrity, architectural shifts in term of layouts and the arrangement of framing and stiffening occurred as designers found better ways of using materials. For instance, a major change was the move to longitudinal framing introduced by Joseph Isherwood in 1908. Over time, these ideas have been identified as foundational principles for modern ship designs.[4] Other waves of architectural innovations in this field involve segmentation of features that have created specialized ships such as a Roll On–Roll Off (RORO) designs, where containers and cars could be driven into the decks, development of environmentally efficient fleets, and integration with broader supply chain and logistics capabilities through the usage of information and communication technologies.[5]

Creating Vision

The opportunity for ushering such long lasting cycle of innovations can be usually traced to butterflies that were created by the vision of a single engineer or an architect. One of the pioneer engineers who helped usher this maritime era was Isambard Kingdom Brunel (1806–1859). Brunel has been attributed with the creation of the Great Western Railway that connected London with south west and west of England and most of Wales, followed by a series of famous steamships including the first propeller-driven transatlantic steamship. The Great Exhibition of 1851 in London publicized America's wealth and natural resources and created momentum for emigrating from Britain to America. Brunel recognized the potential for larger ships

---

[3] For the references to the history at the Lloyd's Register of Shipping, see:

• Martin, F.: The History of Lloyd's and of Marine Insurance in Great Britain. Adamant Media Corporation, Boston (2005).

[4] Comstock, J.P.: Principles of Naval Architecture. Society of Naval Architects and Marine Engineers, New York (1967).

[5] http://imtech.eu/EN/corporate/About-Imtech/Visions/Vision-Green-Ships.html.

purpose-built to carry emigrants. He came up with the vision for the Great Eastern, a steamship whose measurements were six times larger by volume than any existing ship. Brunel realized that this ship would need more than one propulsion system: paddles and a regular propeller (known as a "screw" in marine terminology). Deploying paddle wheels meant that the ship would be able to reach Calcutta, a major maritime destination in that era, where the Hooghly River was too shallow for screws. Since twin screws were still very much experimental at that point in time, he settled on a combination of a single screw and paddle wheels, with auxiliary sail power. At its launch, the Great Eastern was the largest ship built and had the capacity to carry 4,000 passengers around the world without refueling. She plied for several years as a passenger liner between Britain and America, before being converted to a cable-laying ship for the first lasting transatlantic telegraph cable in 1866.[6] The Great Eastern finishing her life as a floating music hall and was broken up in 1889.

Designing for Adaptability: Platforms and Modular Options

One might argue that the success of the Great Eastern is owed both to the vision of its architect and the ability of its users to adapt it for different business applications. Adaptability can be informed by moving away from the realm forecasting into the realm of scenario planning. Since the focus of our chapter is on leadership, we refrain from offering a technical analysis of the mechanics of developing more flexible and adaptable architectures in terms of platforms and modules—a supplement on this area is provided in the appendix. A central idea in such settings is modular decomposition of complex tasks or artifacts into simpler subtasks or artifacts. These ideas have been well understood in complexity science following the arguments put forth by Herbert Simon (1962). Simon described in detail a process of decomposition of a project into subprojects that enabled parallel and/or specialized effort by different parts of the organization, which were then followed by integration of the parts into a whole.[7] However, perfectly decomposing a project is rare owing to a variety of problems ranging from underlying physical laws relevant to the project to cognitive limits of the innovation workers to the principle of providential behavior that creates distortions across organizational boundaries. Instead, the subprojects are typically designed into a nearly decomposable system that comes with some information dependence across interfaces. Hence, good architects pay explicit attention to modularity during development, production (if applicable), and

---

[6] For the history and impact of the transatlantic cable, see:

- Murray, D.: How cables unite the World. In: The World's Work: A History of Our Time: 2298–2309. A Google Book) (1902).
- Dibner, B.: The Atlantic Cable. http://www.sil.si.edu/digitalcollections/hst/atlantic-cable/ac-index.htm (1957).

[7] Simon, H.A.: The architecture of complexity. Proc. Am. Philos. Soc. **106**(6), 467–482 (1962).

customer use while organizing their tasks.[8] It is worth noting that modularity in development seeks to minimize the complexity associated with development, when separate groups or individuals work on two aspects of a development project. The goal is usually to minimize the schedule (and cost) while meeting a shared performance goal. Modularity in production, on the other hand, seeks to enhance the efficiency of the production process by decomposing the pieces what may work upon by separate entities in a supply chain. Modularity during customer use deals with another goal—to create decomposition between parts that will allow ease of use, replacement and/or substitution. Given these disparate goals, it is rare for decentralized teams to achieve even near-decompositions on all three types of modularity, and thus innovation processes invariably lead to emergent outcomes through the principle of exchange. A question that often comes up, is are there specific leadership actions that can help an innovator in simplifying such evolutionary complexity through modular actions? We address this question in the next two subsections.

Dealing with Unavoidable Interdependence

If interdependence cannot be eliminated, there seem to be three types of actions that architect can seek: speed up the SROM cycle, build in some hierarchy, or use outside agents who can offer entire chunks of readymade functionality with a high degree of reliability. We go back to some examples from the shipbuilding industry, and a few others from more general settings, to think about how these actions can play out. Rapid production of warships and transport ships was at a premium at the onset of the Second World War. The U.S. shipbuilding industry had started producing a Liberty class ships in 242 days on average at the beginning of the war. The industry was able to cut down this number to 42 days by going to assembly line system that deployed prefabricated units. A second way to cut down on the development time is by cutting out some of the functionality, and thus the scope of the tasks, involved in each building project. This may not be possible in some systems, such as the construction of high-rise buildings, which must be done in a particular sequence. However, in other settings, such as software development, it is possible to

---

[8] There is a growing literature on usage of modularity to inform product family and platform designs. See:

- Clark, K.B., Baldwin, C.Y.: Design Rules. Vol. 1: The Power of Modularity. MIT Press (2000).
- Krishnan, V., Ramachandran, K.: Economic models of product family design and development. Chapter 4. In: Loch, C., Kavadias, S. (eds.) Handbook of New Product Development Management. Butterworth-Heineman, Oxford (2008).

Recent research on system engineering has brought in complexity science into the field that is broadly termed as Design for Adaptability (DFAD). We refer the reader, particularly the systems engineers, to the work of our colleague Professor Tyson Browning that lays out a structured methodology:

- Engel, A., Browning, T.R.: Designing systems for adaptability by means of architecture options. Sys. Eng. **11**(2), 125–146 (2008).

arbitrage the features and cut down the cycle time, as described in the discussion of agile software development in Chap. 5.

Hierarchy provides a well-understood mechanism for modularizing a complex organization (such as military "corps" structure pioneered during the Napoleonic period or the divisional organization structure implemented by Alfred Sloan at General Motors) into standalone groups, some of whom are endowed with specialized knowledge. Recent research on hierarchy in new product development settings suggests that the presence of hierarchy can dampen the emergent churn, i.e., escalation and runaway outcomes in projects.[9] However, such modularization can also slow down the SROM cycle by diminishing the need to communicate with higher levels of management. As described in the discussion of agile management in Chap. 5, one way to overcome slow feedback in organizational hierarchies is by embedding managerial representatives, who are not the project managers, into project teams who act in information gathering and feedback mechanisms that provide rapid access to higher-level decision makers. Within new product development, innovation executives and architects can set up such communication channels by establishing personal contacts with a few selected technical leaders. The motivation for these technical leaders is the enhanced likelihood of harnessing butterfly effects and peer respect, rather than economic gains. These leaders are thus given access to the big picture by the organization's architects and can therefore decide when relevant emergent information must be gathered and made visible to the senior management through the directed telescopes.

A recent trend has been to manage complexity and reduce interdependence by going away from hierarchical organizations and instead search the marketplace for entire chunks of ready-made innovation. Examples of such market places are open innovation and innovation tournaments.[10] Open innovation focuses on the use of purposive inflows and outflows of knowledge to accelerate internal innovation. Tournaments expand the markets for external use of innovation. However, it is critical that firms that buy these modules define and advertise clear interfaces. It is also important that the "integrators" can quickly evaluate and assemble these chunks and stitch together a "network of commitments" to support them.[11] The role of the

---

[9] Mihm, J., Loch, C.H.: Spiraling out of Control: Problem-Solving Dynamics in Complex Distributed Engineering Projects. In: Braha, D., Minai, A., Bar-Yam, Y. (eds.) Complex Engineering Systems. Perseus Books, Springer, New York (2006).

[10] For discussion of open innovation see Chesbrough (2003). Innovation tournaments are described by Terwiesch and Ulrich (2009).

- Chesbrough, H.: Open Innovation: The New Imperative for Creating and Profiting from Technology. Harvard Business School Press (2003).
- Terwiesch, C., Ulrich, K.: Innovation Tournaments: Creating and Selecting Exceptional Opportunities. Harvard Business School Press (2009).

[11] For additional discussion such process centric work, see:

- Sull, D.N., Spinosa, C.: Promise-based management: the essence of execution. Harv. Bus. Rev. 85(4), 78–89 (2007).
- Lévárdy, V., Browning, T.R.: An adaptive process model to support product development project management. IEEE Trans. Eng. Manage. 56(4), 600–620 (2009).

architect in such settings involves estimating how much interaction and coordination burdens that these options can alleviate and balance these gains against the additional issues that these decisions are likely to create based on the principle of exchange.[12]

Understanding and Accounting for Stakeholder Aspirations

The principle of providential behavior suggests that differences in aspirations between stakeholders will pull the outcomes of any decentralized project in different directions. It is incumbent upon the architect to understand and think through the goals and aspirations that drive a diverse set of stakeholders. One way to account for such aspirations is by asking these stakeholders about their perception of the impact of a local decision, largely under their own control, on the global outcome that will emerge though the interaction between different stakeholders. In certain domains such as software engineering, architects now create simulations using methodologies such as the unified modeling languages (e.g., UML©), to inform the stakeholder about the possible interactions before asking for their assessment of a complex outcome.[13] Research on these experiments in the innovation setting seems to support the notion that stakeholders will bias their choices in accordance with their own aspirations, as one would expect from the principle of providential behavior. For instance, when asked about the anticipated level of project progress, design engineers thought that their portion of the task, the front end, would evolve slowly compared with the back end, while a test engineer who would shoulder much of the tasks at the back end argued for a reverse pattern.[14]

Aside for these estimations, a key idea to keep in mind is that the action of an architect in an innovation-driven organization symbolizes that organization's hopes for as better future. Frank Lloyd Wright described the importance of the day-to-day decisions made by an architect: *the present is the ever moving shadow that divides yesterday from tomorrow. In that lies hope.*[15]

Leveraging Information Technology and Analytics

Given that architects actions must be based on a keen understanding of their organization behavioral biases, relevant technologies, and market needs, it is incumbent upon an architect to gather and process huge amounts of information. This cannot

---

[12] Anderson, E.G., Davis-Blake, A., Parker, G.G.: Organizational Design for Outsourcing Complex Tasks. University of Texas Working Paper (2010).

[13] Joglekar, N.R., Yassine, A., Eppinger, S.D., Whitney, D.E.: Performance of coupled product development activities with a deadline. Manage. Sci. **47**(12), 1605–1620 (2001).

[14] Ford, D.N., Sterman, J.D.: Expert knowledge elicitation to improve formal and mental models. Syst. Dynam. Rev. (1996).

[15] Wright, F.L.: The Living City. Horizon Press (1958).

be done without deploying information and communication technologies (ICT) in a thoughtful manner.

We draw upon a vignette provided by Bill Gates about the manner in which he worked while he was the Chairman and the Chief Software Architect at Microsoft.[16] Gates describes his usage of e-mail systems and a collaboration tool called SharePoint. At Microsoft, e-mail was the key medium of communication, more than phone calls, documents, blogs, bulletin boards, or even meetings. Part of this is due to the fact that voicemails and faxes are integrated into Microsoft's e-mail in-boxes. Gates indicated that he read about 100 emails every day. The mail was filtered in a manner so that e-mail came straight to him from anyone whom he ever corresponded with, anyone from Microsoft, Intel, HP, and all their other partner companies, or anyone he knew. In effect, these messages came from people who enabled him to scout the market landscape in a directed manner. In addition, the usage of SharePoint provided a quick access to websites for collaboration on specific projects. These sites contain plans, schedules, discussion boards, and other information, and they can be created by just about anyone in the company.

A second key dimension of this information processing system seems to be the manner in which one communicates aspirations and gathers feedback. Gates stated, "Microsoft has more than 50,000 people, so when I'm thinking, 'Hey, what's the future of the online payment system?' or 'What's a great way to keep track of your memories of your kid?' or any neat new thing, I write it down. Then people can see it and say, 'No, you're wrong' or 'Did you know about this work being done at such-and-such a place?'" Such documentation of the architectural vision, or open ended statements, leaves room for technology savvy subordinates or peers to comment on potential plans, before they get set firmed up into formal projects. It also fosters organizational learning through scenario planning and dialogue.

Summary

Unlike Bill Gate's position at Microsoft, most organizations do not have leaders who identify themselves as architects. However, any innovation leader, be it a CEO, a VP of Marketing, R&D Director, or an individual contributor in a specialized domain must spend a considerable time in architecting the organizational structures that shape their teams' innovation offerings. Furthermore, they must develop an understanding of interrelationships among their key parts of this structure and set up the organizational processes that support it. To facilitate this, we have identified key actions ranging from focusing on scientific advances to the need to leverage information and communications technologies that innovation leaders must use to facilitate their architecture. Leaders who ignore their role as architect risk creating an innovation organization that cannot harness the innovation butterfly, but rather must lie at its mercy.

---

[16] Gates, W.: How I Work, Fortune Magazine (2006).

Chapter 9
Leader as a Ship's Captain

To Go as far as a Man Can Go

My ambition is to go not only farther than any man has been before me, but as far as I think it is possible for a man to go.

Captain James Cook[1]

As discussed in the previous chapter, Peter Senge in the *Fifth Discipline* pointed out that the most "powerful" position on a ship might not be the captain, but rather the ship's architect. While this view is indeed useful, in the realm of distributed innova tion leadership, captaining the ship is a critical piece of the leadership puzzle as well. Otherwise, an innovation leader with poor captaining skills will soon ground her ship upon the shoals of the three principles of escalation of expectations, exchange, and providential behavior.

Because we are discussing many types of leadership roles within the world of product and service development, with all of its unknowns as magnified by the principle escalation of expectations, the particular type of captain we would imag ine that best exemplifies the skills needed by the innovation leader would be the great explorers such as James Cook, the famous British sea captain.[2,3] We highlight several aspects of the role of a leader as a captain: orchestration through discretion in command, planning for the unknown, scouting, managing personnel, and lever aging new technology.

---

[1] http://www.bbc.co.uk/history/british/empire_seapower/captaincook_01.shtml.

[2] Shatner, W.: Up Till Now: The Autobiography. Macmillan (2008).

[3] Fisher, R., Johnston, H.: Captain James Cook and his Times. Taylor & Francis. pp. 81, 86, 90, 96 (1979).

E.G. Anderson and N.R. Joglekar, *The Innovation Butterfly*,
Understanding Complex Systems, DOI 10.1007/978-1-4614-3131-2_9,
© NECSI Cambridge/Massachusetts 2012

Orchestration Through Discretion in Command

There exist a number of enlightening parallels between Cook's three voyages of exploration and the work of leaders in a decentralized (or distributed) innovation setting, particularly for any innovation leader located far away from headquarters and charged with a roadmap that must be executed.[4] Cook was given immense discretion in his command. He had to have this because he would be out of communication with sponsors in London, the British Admiralty, literally for years on all three voyages and the likelihood of encountering any other vessels, either British or foreign, would be essentially zero for long stretches. Accordingly, the Admiralty kept his orders extremely simple and straightforward. For example, for his first voyage, his mission was to create a scientific record of the transit of the planet Venus across the Sun in 1769. The voyage, once the transit had been recorded, was to have a second purpose, which was to probe for the existence of any substantial unknown landmasses to the southeast of Tahiti. Much of how he was to accomplish these goals, however, was left to his discretion, because he would be the person with the best and most up-to-date knowledge available on how to accomplish his mission. This was not unusual. The British Admiralty had learnt the hard way through many misadventures to avoid micromanaging individual ships when they were away at sea.[5] Because the Admiralty could not understand their individual commanders' circumstances in detail, any overly detailed instructions would restrict the actions of those commanders, reducing their effectiveness in dealing with complex phenomena. For example, detailing how Captain Cook should engage with newly encountered indigenous peoples could not be easily imagined beforehand, because the behavior of the Tahitians and the Australian Aborigines were as different from each other as they both were from that of Europeans.[6]

Similarly, innovation executives in today's distributed settings (including offshoring and outsourcing) are often located far away from the site where problems arise and get solved. While providing a roadmap to their subordinate teams and orchestrating their high-level behavior, off-site executives should avoid attempting to over-specify their team's behavior while maneuvering. Otherwise, because of the "fog" in the innovation system described in Part II, problem solvers "on the ground" will find their hands tied as they improvise in response to the complexities of the innovation system.

---

[4] Unless state otherwise, the discussion of Captain Cook's exploits and the policies of the British Admiralty in this chapter are drawn from:

- Fisher, R., Johnston, H.: Captain James Cook and his Times. Taylor & Francis. pp. 81, 86, 90, 96 (1979).
- Hough, R.: Captain James Cook: A Biography. Norton, New York (1994).

[5] Rodger, N.A.M.: The Command of the Ocean: A Naval History of Britain, 1649–1815. Norton, New York (2005).

[6] Hough, R.: Captain James Cook: A Biography. Norton, New York (1994).

Planning for the Unknown

Captain Cook also put painstaking detail into preparing for his expeditions, which explains the high survival rate of his crew across multiple voyages. Cook was selected as commander of his first voyage too late to influence the selection of his ship for that voyage, the *Endeavor*, but he learned from this experience. On his later voyages, he selected ships that were similar to *Endeavor*, which was a merchant ship able to sail well both in shallow waters as well as in the deep ocean. Another organizing principle was that there were a number of contingencies in which two ships would prove much more capable than one larger ship. In the worst case, for example, it was highly unlikely that two ships would be simultaneously lost. Thus, if one ship were lost, the second ship would provide some margin of safety for any survivors. He was also willing to test and adopt new technology such as the chronometer method of measuring longitude, which could simplify and improve his navigation, which in voyages to unknown waters might mean the difference between life and death. It is important to note, however, that Cook still had the backup method of determining longitude based upon astronomical observations that was known to work well enough if the chronometer method failed. Both of these examples have parallels in the work of the innovation leader in that planning is crucial and that the modularization of risk, including technological risk, is an integral part of that planning.

Scouting

Another instructive parallel was that Cook was an outstanding navigator, surveyor, and cartographer. Early in his career, he was charged with making the charts that made is possible for the capture of French Canada by General Wolfe's famous amphibious expedition up the St. Lawrence River to capture Quebec City in 1759. Later, Cook surveyed much of the rocky, treacherous coast of Newfoundland, which was of strategic importance to the European economy of that time due to its immense stock of fisheries. In fact, this cartographic ability was one of the key reasons he was chosen by the British Admiralty to command his first voyage of exploration. This paid off handsomely when Cook became the first person to chart the east coast of Australia as well as virtually all of New Zealand. These charts proved to be extremely accurate to the extent that some of them remained in use a century and a half later. Furthermore, beginning with Tahiti, Cook included at the end of his journal a summary description of each land he visited, including not only good anchorages and approaches, but also whether firewood was easily obtainable, the nature of any potential trade goods, and a description of the peoples encountered. Perhaps more importantly, he noted what he did *not* know, such as the religion of the Tahitians, which "is a thing I have learnt so little of that I hardly dare touch upon it." Components of scouting, such as tracking and charting, are key to developing an understanding of the evolution of innovation projects and portfolios. We describe some of the relevant analytical capabilities in the appendix.

It is worth noting that Captain Cook included such speculative musings as his belief that the various Polynesian peoples were related and that they originated from Asia. All of this became of immense value to the Admiralty because the point of all voyages of exploration was to build up institutional knowledge that could be later disseminated in future endeavors. If Cook had simply made the voyages and only left poor charts and descriptions of where he visited, the voyages would have been of much less value. Similarly, it is important to realize that the work of the innovation leader is not only the execution of an innovation project or projects, but also to learn from these projects and to disseminate that learning to the rest of the organization through careful after action reviews.[7] This can indeed be done by other individuals within the innovation team such as the management representative. However, the project leaders must facilitate and support this work, because the documentation of the nature of that portion of the innovation system encountered is the basis for scouting during the next SROM cycle and ultimately the basis for organizational learning in an innovation firm.

Personnel Selection and Oversight

Cook gave much thought to the selection and oversight of his crew. For the crew of the first voyage, Cook selected many of his subordinates from those who had served with him before in Newfoundland. While it was typical to bring along people from previous commands at the time, Cook also no doubt recognized that they brought a rare measure of experience in geographic exploration as well as contact with indigenous peoples. Cook increased the corporate experience of his crew by recruiting six veteran sailors who had already circumnavigated the globe during voyages of exploration—an extreme rarity for a sailor at the time. These new crewmembers also brought first-hand experience of Tahiti, where Cook would have to make landfall to reprovision before exploring to the south.

In particular, while he was known as being relatively—compared with the draconian practice of the time—lenient on discipline, he was quite strict on enforcing shipboard cleanliness and a diet laced with such scurvy preventatives such as sauerkraut in an effort to preserve the health of his crew. Even so, at first the crew on his first voyage refused to eat sauerkraut. According to Cook's journals, he solved this problem by at first ordering it only to be served to himself and to his officers, leaving the sauerkraut as an option for any other crew member who wished to have it. Soon enough the crew was eating so much sauerkraut that it had to be rationed. Similarly, Cook put a premium on dividing food as evenly as possible among all members of the crew. The lowest seaman ate the same rations as the Captain and his lieutenants. Besides the obvious morale implications, it also enabled Cook to

---

[7] Darling, M., Perry, C.: From Post-Mortem to Living Practice: An In-Depth Study of the After Action Review. Signet Research and Consulting, L.L.C., Boston (2002).

distribute all the fresh and exotic food that they could find when reprovisioning, reasoning that a varied diet would improve the health of the crew. A particular testament to the capability of Cook's management of his crew was the absence of any serious threat of mutiny during his three voyages. The threat of mutiny seemed to be an endemic problem for voyages of exploration ranging from Magellan and Hudson down through Cook's own protégé, William Bligh, when he later became commander of the infamous *HMS Bounty*.

The parallels of Cook's approach to managing his crew for the innovation leader are twofold. One is to select the right skill-set for the personnel of a project. Like Cook's selection of sailors with experience of circumnavigation and Tahiti, related experience in technologies or markets is extremely beneficial, particularly among the lower level leaders of a project, to whom most of the detailed problem-solving assignments are likely to be delegated. Of equal importance, however, is need for these leaders to understand how to motivate a project's personnel. While innovation workers are not sailors, as Paul Glen, David Maister, and Warren Bennis in *Leading Geeks: How to Manage and Lead the People Who Deliver Technology* teach us, they are also not typical corporate employees.[8] Learning how to motivate them in ways other than through money or promotion (or fear)—none of which work well with innovation personnel—is essential. In particular, innovation workers seem to respond well—according to Glen, Maister, and Bennis—to recognition of their capabilities and the opportunity to take on more difficult challenges.[9] However, the details are almost unimportant. The point is that many of these personnel (1) will know many valuable things that the innovation leader doesn't, particularly with respect to technology, and (2) are often driven by motivations different from management. Learning what these motivations are and then using them appropriately, like the issuance of sauerkraut to Cook's officers to induce its eating by the common sailors, is the key. Otherwise the malign side of the principle of providential behavior will arise and the result will be a dysfunctional team that can wreck an innovation project, even when all other indications are that the project should succeed.

Improvisation Through Delegation

Perhaps the most interesting and relevant illumination of Cook's management style was how he handled matters when things did not work out as expected. The most perilous portion of Cook's first voyage was when the *Endeavor* struck the Great Barrier Reef of Australia. It took the *Endeavor's* company over 24 hours to get off the Reef. According to members of the crew, part of what made that incredibly stressful period bearable was the calm composure of Captain Cook. However, Cook's behavior becomes even more interesting when the crew finally managed to

---

[8] Glen, P., Maister, D.H., Bennis, W.G.: Leading Geeks: How to Manage and Lead the People Who Deliver Technology. Jossey-Bass (2002).

[9] Ibid.

get the *Endeavor* off the reef, and found that it had begun to take on so much water that the pumps could not keep up. In short, the ship was sinking. At this point, Cook's care in selecting the crew paid off. One of the midshipmen, Jonathon Monkhouse, suggested using a technique that he had learned in a prior merchant voyage across the Atlantic in a leaky ship for using a sail to temporarily plug the leak from beneath the hull. Once Cook accepted his suggestion, he left Monkhouse to direct the execution of this difficult repair without any further interference. Later, while repairing the *Endeavor,* he and his crew took advantage of the month on land to expand their description of the wildlife of Australia, including the most famous of its naturalists' findings, the first scientific description of the Kangaroo.

There are a number of insights in Cook's handling of these incidents for innovation leaders. One is that much of the innovation leader's time will be taken up by unexpected events and their aftermaths, i.e., the innovation butterfly's effects. Leaders must be prepared for coping with them, and the best coping mechanism is not a tantrum, but a team of capable individuals with deep experience. Another lesson is the fractal nature of delegation. Much like the Admiralty's avoidance of micromanaging Captain Cook, Captain Cook realized that Monkhouse knew much better how to deal with the leak than Cook did, and so he left Monkhouse to deal with it without interference. A leader could exercise foresight, and mentor potential Monkhouses on the teams. A final insight is for innovation leaders to try to take advantage of innovation butterflies, even unfavorable ones, in order to further the progress of the firm as a whole through increasing its organizational learning.

Impact of Information Technology

Obviously, many things have changed since Cook's day. For example, centralized maritime command authorities, such as the British Admiralty, can communicate much more frequently and easily with their subordinate commanders, no matter where they are distributed around the globe. However, despite improved communication, which effectively shortens their command and control cycle, much discretion is left with ships' captains as illustrated by the recent attempted hijacking of the U.S. merchant ship Maersk-Alabama.[10] Its captain, Richard Phillips ignored an advisory from the day before to remain at least 600 miles off the coast of Somalia, apparently because remaining that far offshore would have lengthened their sailing time by over a day.[11] During the hijacking, pirates boarded the Maersk-Alabama

---

[10] The description of the Maersk-Alabama Hijacking has been subject to a great deal of controversy. See:

- Phillips, R., Talty, S.: A Captain' Duty: Somali Pirates, Navy SEALS, and Dangerous Days at Sea. Hyperion, New York (2010).
- Payne, J.C.: Piracy Today, Fighting Villainy Today. Sheridan House, Dobbs Ferry, NY (2010).

[11] Curran J.: Mutiny: Crew Blames Richard Phillips, Maersk Alabama Captain, For Ignoring Pirate Warnings. www.huffingtonpost.com/2009/12/03 (2009).

280 miles south east of the port of Eyl in Somalia, and, when they failed to take the ship, removed Captain Phillips to their speedboat and held him for ransom. While Phillips' decision to ignore the 600-mile warning may appear poor in retrospect, his rescuer, the USS Bainbridge was under similarly decentralized control. While its captain, Frank Castellano, was in contact with President Obama during the hijacking, the President merely affirmed standing orders to attack the pirates with deadly force if Castellano judged Phillips to be in imminent danger and left such judgment to Castellano's discretion. Castellano, in turn, left the actual execution of the raid to U.S. Navy Seal Snipers parachuted in for the emergency.[12]

In essence, while the tempo of the command and control loop cycle increased, and many modern day information and communication technologies are available to teams, the decentralization of command is still necessary. It is difficult to imagine that the President or some other off-site functionary could have provided real-time oversight to the SEAL Snipers with the speed and flexibility required by the situation. Similarly, modern transportation technology permitted the transfer of specialized personnel to the *Bainbridge* at a speed that would have been unimaginable in Cook's day. However, even if they were unavailable, it is hard to imagine that Captain Castellano would not have delegated the raid's execution to the security forces within his crew, much like Captain Cook left the repair of the *Endeavor* to Midshipman Monkhouse.

For innovation leaders, the lesson should be analogous. Improved information technology permits a faster SROM tempo and must be exploited to permit more frequent orchestration. That said, it does not obviate the necessity for delegation during the maneuver phase of operations to personnel on the spot and, perhaps even more importantly, avoiding getting out of their way afterward.

Summary

In short, the description of various challenges faced by Captain Cook, and the manner in which he prepared for these challenges and adapted to the situations, provides a number of rich ideas for the innovation manager to ponder. The first and most important is the need for higher level innovation leaders to provide their subordinate innovation leaders with a clear understanding of the objectives and intents behind a maneuver during the SROM cycle, but then keep out of the way of the subordinate leaders during the execution of the maneuver. Another key idea is the need at all times to document the details associated with the execution of an innovation project so as to increase organizational learning about the innovation system it is embedded in. A third idea is the necessity for meticulous planning as the first part of a maneuver, particularly in modularizing risk, but then once that maneuver is

---

[12] Associate Press (2009) WRAPUP 10-U.S. Navy rescues captain, kills Somali pirates, retrieved on 2 July 2010.

being executed, to be prepared for improvisation as the effects of innovation butterflies arise. Fourth is the importance of selecting personnel with the correct skill-sets for the innovation project and learning how to motivate them. Finally, the necessity for subordinate innovation leaders to realize that they are in fact managing their own SROM cycle and, hence, must also exploit their subordinate's capabilities but then stay out of the way of those subordinates, once they begin to execute their own innovation work.

Chapter 10
Leader as a Coach

It's not About Talent, It's About How the Team Plays

> *It's not about talent, it's about how the team plays.*
> *That's the litmus. Let's see how we play,*
> *Let's see how we coach.*
>
> Bill Belichick[1]

So far we have discussed the parallels between the innovation leader and architects and ships' captains. Now, to complete our discussion, we examine the parallels between the innovation leader and a sports team's coach. We do this because of the need to create a culture for the innovation firm that nurtures maneuver-driven competition while avoiding the worst aspects of the principle of providential behavior. Doing so requires the nurturing of a set of norms and values that build upon a shared "love-of-problem solving," yet places the success of the firm as a whole above that of its component teams or individual members. It is no coincidence that individuals who display such behavior are often referred to as "team-players" or "good sports." The earliest experience most of us have with working in teams occurs within the venue of sports, and many of us follow the exploits of our favorite teams through good times and bad for the rest of our lives. Many authors of organizational science have effectively used sports metaphors to illustrate organizational truths. For example, Peter Senge illustrated his discussion of effective team learning in *the Fifth Discipline* with examples from Basketball.

We will follow Senge's lead but instead examine the factors underlying the cultures of successful teams in America's National Football League (NFL). We do this because the task of coaching an NFL football team bears strong resemblances to leading an innovation firm. An NFL head coach must manage a roster of 53 highly specialized players, including such oddities as the "long snapper," whose peculiar job is to replace the center offensive lineman and snap the football—i.e., toss it backward through his legs at the start of a play—up to 15 yards in 0.7 second or less

---

[1] http://www.patriotsbook.com/volume1.html.

E.G. Anderson and N.R. Joglekar, *The Innovation Butterfly*,
Understanding Complex Systems, DOI 10.1007/978-1-4614-3131-2_10,
© NECSI Cambridge/Massachusetts 2012

during plays involving either kicking or punting. Eleven of these variegated specialists are on the field at one time, and substitution can occur prior to each play—thus, success depends greatly on the orchestration, i.e., selection by the head coach as to which set of 11 out of 53 players are asked to be on the field for a given play and also the playing instructions provided to them by positional coaches. The head coach must ensure that a team has strong individual players in a number of different specialties who can meld their play together to execute the correct game plan. Furthermore, the head coach must do so in the certainty that things will not go as expected. Over a season, balls will bounce one way or another in any given game, key official rulings will be misjudged, and some of the players will be injured. These disruptions must be managed in order for a team to achieve success. Moreover, if a team excels over time, other teams will learn from its success and adapt to either copy or counter balance any successful strategies.

For example, Bill Walsh, who was the head coach of the three-time Super Bowl winning San Francisco 49ers, popularized a version of the so-called "West Coast Offense" strategy that relied on extremely predictable short passes from the quarterback to his receivers to advance the ball rather than handing off the ball to one of his running backs.[2] This caused the players to spread from the center of the field to its edges on both sides of the field as well as other ramifications that are still playing out to this day. For example, according to Michael Lewis, author of *The Blind Side: The Evolution of a Game,* the West Coast Offense strategy increased the effectiveness of the quarterback and, in turn, led to changes in related defensive strategies (such as the "3–4 defense," specifically designed to put pressure on the quarterback). Over time, this evolution again led to changes in the West Coast Offense-like strategies, including the focus on finding protection for the quarterback in the person of especially gifted left offensive tackles, the player specifically assigned to protect the quarterback from tackles coming from his left-hand side, which the quarterback is effectively blind to while throwing. This put a premium on finding such suitable left-offensive tackles and resulted in becoming the second-most highly paid player on a team after the quarterback.[3] Furthermore, the 49ers success with the "West Coast Offense" caused it and its collateral relative, the "spread offense," to be adopted by numerous teams such as Bill Belichick's New England Patriots, which over the past decade has won three Super Bowls of its own.[4] If one could change the names of the teams to innovation firms and the names of the offenses to innovative products, one could hardly ask for a more classic example of the principle of exchange in an innovation system.

---

[2] The information concerning Bill Walsh's career, unless otherwise stated, comes from

- Harris, D.: The Genius: How Bill Walsh Reinvented Football and Created an NFL Dynasty. Random House, New York (2008).

[3] Lewis, M.: The Blind Side: Evolution of a Game. Norton, New York (2006).

[4] Information regarding Bill Belichick's career, unless otherwise stated, come from

- Halberstam, D.: The Education of a Coach. Hyperion, New York (2005).

Capability Planning and Modularizing Risk

Not surprisingly, there are many similarities between a successful football head coach and a successful innovation leader. Capability planning and development is the key to football as well as innovation. No coach can be successful without developing a team with strength in a number of very different capabilities, both at the individual and collaborative levels. For example, while no team can survive long without a skilled quarterback, that quarterback's effectiveness hinges on many other capabilities such as that of the offensive linemen to block, the running backs to rush, and the receivers' ability to catch. And, of course, each team member's individual skills, such as speed, strength, blocking, and tackling, must be developed as well. However, all of these skills and capabilities must be matched to the overall strategic philosophy of the team and along with their ability to work well together. It is this team spirit or sense of cooperation that gets teams across the finish line. For example, some teams prefer to primarily rely on a deep passing game and less on running. This has certain personnel ramifications, such as additional wide receivers or tight ends—whose job is to catch the ball—because a deep passing game implies that receivers and tight ends will be on the field more of the game than for other sorts of offensive schemes. Furthermore, developing a starting lineup is not enough to guarantee success. The number of injuries and simple wear and tear inherent in a full season of football requires in-depth skills in every position. Additionally, because each week's game brings a different opponent, in every game the team must develop a different game plan and hence rely on a different set of player capabilities. Acquiring new players takes time and money, and developing players' skills incrementally also takes time. This is where the ideas such as modular management of capability portfolios can be put into practice (e.g., some players make the roster as a generalist because they contribute in limited roles as defensive back and also as a special team player. Over time, these players are given the opportunity to develop their capabilities for more skilled positions, e.g., such as a starting defensive back). The result is that a new coach often needs several years to assemble a team with the capabilities to match his strategic philosophy, particularly with enough depth to protect against the inevitable uncertainties of the game over a season.

In recent years, similar to innovation executives, coaches' capabilities have been enhanced by the use of information systems to analyze statistical data.[5] Head coaches are also assisted in each game by extensive analysis of the opponent's strategy and capabilities in earlier games by advanced scouts and film analysts.[6] Interestingly, Bill Belichick's New England Patriots, who are considered one of the

---

[5] Davenport, T.H., Harris, J.G.: Competing on Analytics: The New Science of Winning. Harvard Business School Press, Boston (2007).

[6] Halberstam (2005).

most analytically advanced franchises in the league, find that the biggest benefit of this information is to manage their personnel and assess their capabilities.[7]

Creating a Successful Culture

Although it has less of a lead time, adjusting the strategy and tactics for each particular opponent is equally critical in football, particularly since a team's capabilities may often fall short of the coach's ideal. Fortunately, the head coach need not do this alone. The actual plays run in any game are generally called by a defensive or offensive coordinator (and in some rare cases by a savvy quarterback such as Peyton Manning). Conflicts and lack of communication between the head coach and the coordinators over play calling and other issues typically lead to confusion on the field, and result in poor performance. Hence, football coaches must, in most essentials, manage a complex environment using an SROM cycle of maneuver-driven competition by building agile capabilities and modularizing risk in a manner similar to an innovation leader. However, this is not enough to enable a truly successful team over the long run, and this is where the role of the head coach becomes most illustrative. Most successful coaches are known for trying to develop a certain character or ethos on their team. In some instances, because the celebrity status and money that go along with playing in the NFL (both for the player and their coaches), tends to promote self-centered behavior, even if it is detrimental to the team, thus kicking the principle of providential behavior into overdrive. The behavior of wide receiver Terrell Owens of the San Francisco 49ers, Philadelphia Eagles, Dallas Cowboys, and Buffalo Bills is one notorious example of a player, who despite outstanding skills and performance on the field, was let go by the Bills, the Cowboys, and the Eagles.

The process of building a collaborative and winning culture begins by recruiting or promoting players who are willing to buy into a "team first" philosophy and conduct themselves accordingly. For example, New England Patriots' head coach Bill Belichick is known for looking for players with "no ego," a willingness to work incredibly hard, and an uncompromising love of the game.[8] This creates a virtuous cycle because this sort of player over time often becomes the leader of the team, training younger players to fall into the same mode of conduct.[9] Bill Walsh once stated that "the critical factor whenever people work together is that … the players expect a lot of each other."[10] He went even further by actively recruiting personnel like veteran Jack "Hacksaw" Reynolds in 1981, who could inspire his teammates to

[7] Sauser, B.: Analytics in football. Technol. Rev., February Issue (2008).
[8] Halberstam (2005), Chapter 12.
[9] Halberstam (2008), Chapter 13.
[10] Harris (2008), p. 92.

strive for this leadership spirit. "Hacksaw did a lot for this team. He created a lot of good habits for this organization," according to Ronnie Lott—a superstar player in own right—who played on this team.[11] For example, Reynolds was legendary for obsessively conducting much of his own scouting, or advanced research, on future opponents through watching film of opponents' previous games, more so than many coaches.[12] Innovation leaders should similarly recruit personnel with a "love of problem solving" during R&D along with an ability to work within the spirit of their team. More importantly, innovation leaders should reward, promote, and retain such personnel, if necessary by developing a parallel path of promotion for technical specialists. If anything, the cultivation of superior "Grayhairs" within an innovation team is even more important in innovation firms than on football teams, because the technical expertise and institutional memory of an innovation firm lie much more within middle management and technical specialists than in football, where it tends to reside within the coaching staffs.

Care and Feeding of the Innovation Worker

Another practice that Walsh instituted to mold the culture of the 49ers would probably work just as well in the highly creative world of innovation workers who, as Glen, Maister, and Bennis pointed out, like to be rewarded for good work with even more interesting tasks.[13] Specifically, Walsh made sure that everyone on the team knew that "everyone has a role and every role is essential."[14] Further, his biographer Davis Harris goes on to point out that even if individual players' roles were invisible to spectators, players felt that "if you did your role well, he would recognize you for it. He would design plays for your skills. He always knew which guys plugged in best in which situation. By valuing everybody's role, he made sure nobody checked out and everybody kept their head in the game."[15]

In a further useful parallel, Walsh contributed to this spirit—at least in his early years—by eschewing the then dominant idea that coaches motivated a team by screaming at it and belittling its accomplishments. He believed that it was more efficient if he spoke to his players as if they were respected peers, which they were![16] Again, if anything, screaming and berating highly trained and intelligent workers is even more counterproductive in innovation firms than on football teams, particularly given that—from the authors' experience—innovation workers tend to have extremely long memories.

---

[11] Harris (2008), p. 131.
[12] Harris (2008), p. 130.
[13] Leading Geeks reference.
[14] Harris (2008), p. 89.
[15] Harris (2008), p. 91.
[16] Harris (2008), Chapter 10.

Preparation Breeds Execution and Execution Breeds Success

Walsh made two other vital contributions to team culture in the NFL. One was his mantra that "preparation breeds execution and execution breeds success"[17]; this he believed was more important than players' emotional intensity, which was thought to be a key attribute for ball players at the time. Walsh counted on his players to be intense at all times, and indeed cultivated it, but he also ran endless practices so that his players could execute their prescribed jobs—tackles, running routes, etc.— under any conditions with perfect (or as close to perfect, given the presence of a hostile opposing team) predictability. This endless quest for perfection is what gave life to the short pass of the West Coast Offense, because the quarterback knew exactly how long a receiver would take to get to the end of his running route to catch the ball. So the quarterback, anticipating his destination, could throw to the destination and count on the receiver to be there to catch it. At the time, conventional football wisdom had it that the key to decreasing the risk of losing the control of the ball was to avoid the passing game, because passes traditionally involved a high probability of turnover. Walsh, by radically improving the precision of short passes, effectively invented another way to get the ball to the receiver without increasing the odds of a losing the ball to the other team, thus increasing the number of low risk options open to a team. In a similar manner, increasing the predictability of noninnovative tasks through process improvement can reap significant benefits by reducing the need to modularize risk.

Another key to good team process was the ability to take calculated risks and improving veracity in communication between all team personnel. Walsh once stated, "The critical factor is that when you make a mistake or a miscalculation, admit it. We openly talk when things go poorly and initiate a process to reverse and change the miscalculation."[18] Interestingly, these are precisely the same skills that underpin maneuver-driven competition. To the extent that the innovation leader foregoes telling the truth and cultivating it in her subordinates, the "fog" inherent in innovation systems will necessarily increase leading to poorer execution of the SROM cycle.

Implications for Innovation Management

A football head coach must scout or conduct research on potential recruits, assemble a team of players with key capabilities, develop strategies and tactics, empower them, study the strategies and tactics of opponent teams, and promote a winning or "can do" attitude. Furthermore, all of these elements must work together. The underlying challenges are quite similar to those that an innovation leader faces. The innovation leader

---

[17] Harris (2008), p. 90.
[18] Harris (2008), p. 91.

and her team must do several things. One is to develop key technological capabilities by recruiting and retaining the personnel with key skill sets. Another is to design an appropriate portfolio of product offerings that adjusts to changing market conditions as well as develop processes and tools to enable successful and timely project execution. Finally, and most importantly, they must create a culture that values cooperation and scorns one-upmanship and intrateam competition.

Chapter 11
Epilogue

Chaos in the world brings uneasiness, but it also allows the opportunity for creativity and growth.

Tom Barrett[1]

We began this book with a discussion of the problems inherent in managing innovation systems and some unique solutions for managing them. Leaders within most organizations, whether managing projects, services, manufacturing, or other operations seek to minimize the variability inherent in the system, which in turn improves the ability of the firm to predict outcomes and manage risk. Because the very nature of innovation requires fostering creativity to create a competitive edge, however, innovation leaders find themselves in the odd position of needing to deliberately increase the variability in the system because they need to try out different ideas, which results in less predictable outcomes and greater risk. Moreover, as discussed in the principle of escalation of expectations, innovations are the result of entrenched processes wherein competitive actions, employees' skill sets, interactions among various products in a portfolio, market needs, and investment policies are linked into a complex, multilayered system. Even the smallest change, the smallest disruption, to this system creates a butterfly effect that can steer a firm down an irreversible path in terms of technology and market evolution, and ultimately success or failure. Additionally, other disruptions can evolve from external forces such as government legislation or environmental regulations, or unexpected spikes in the price of oil, and so on, or they can be created by a company manager's decisions or those of its competitors. Hence, the unpredictability created by embracing the uncertainty inherent in a creative system is magnified by the dynamic complexity of the innovation system, making its management a spectacularly difficult challenge for innovation leaders at all levels of the firm.

All innovation leaders, whether senior managers, product line planners, project managers or even technical leads or architects charged with steering a firm's

[1]http://www.brainyquote.com/quotes/quotes/t/tombarrett132437.html.

E.G. Anderson and N.R. Joglekar, *The Innovation Butterfly*,
Understanding Complex Systems, DOI 10.1007/978-1-4614-3131-2_11,
© NECSI Cambridge/Massachusetts 2012

innovation portfolio, must somehow cope with the innovation butterfly and its potential effects. In the majority of cases, innovation butterflies merely result in challenges that take up a significant amount of leadership effort and sap efficiencies within individual innovation projects. However, in the long run, some of them can shift the entire innovation system into unplanned directions. With foresight and the appropriate tools, innovation leaders can successfully anticipate, shape, and address these butterfly effects in a number of ways. We have shown examples from multiple industries that the butterfly effect often present firms with a stark choice: they can either act rapidly to turn these disruptions into opportunities for competitive gains or they can conduct business as usual and be eventually crushed. To convert a disruption into a competitive gain requires an ability to understand the evolution of underlying linkages as well as the design and implementation of an appropriate management structure including processes, strategies, and leadership choices.

We sought to capture the essence of these linkages in terms of three principles: *Escalation of Expectations, Exchange, and Providential Behavior*. By understanding the evolving nature of the innovation system, which we called the *principle of the escalation of expectations*, innovation leaders can make some projections about the expected risks and rewards resulting from a particular company or governmental policy or other change in the system. However, innovation leaders must also figure into their plans two other important factors of innovation cycles. First, any change in the innovation system brought about by their own firm's policies will typically substitute one set of problems for another—although, hopefully, it will be a better set of problems! We refer to this principle in this book as the *principle of exchange*.

The biggest challenge that the principle of exchange poses to innovation systems is that the most effective way to improve the management of complex economic systems, known since the time of Adam Smith, has been to decentralize control and thus empower lower levels of innovation leaders. Doing so allows the firm to develop more complex responses to the demands of managing a complex system. However, because of the *principle of exchange*, there is no free lunch. The problem that arises from decentralization and empowerment is that, because lower-level innovation leaders have a less global view of the innovation system, they will make decisions in part based on their own unique goals and those of their employees, rather than what is necessarily best for the firm as a whole. Furthermore, even if the subordinate leaders do try to work for the firm as a whole, there is every possibility because of different locations, perceptions, and biases, that when their decentralized decisions are taken together, they will not form a coherent whole. The more individual leaders' decisions differ from each other and create a less coherent overall firm response to the innovation system, the more likely that individual innovation leaders decisions will pull the firm in different directions and any resulting butterflies will more likely lead to an industry trajectory that is unstable and harmful to the firm. We describe how these misaligned pulls can arise and damage the firm over the long run as the *principle of providential behavior*.

Understanding these three principles that make the innovation system so fiendishly difficult to manage is necessary. However, such understanding by itself would not suffice for successful handling of the butterfly effects that grow from it. Other actions must be taken. Innovation leaders must carefully plan for the long run and, consider many possible scenarios, some of which are unforeseeable. However, because the business world is in constant flux—as the adage attributed to the philosopher Heraclitus states, "You can never step into the same river twice" [2]—the innovation leader must continually revise those plans, *even before all of the actions from the previous set of plans have played out.* Thus, planning results in a neverending cycle of short maneuvers across the shifting currents, which requires a balance of market forecasting, planning, and execution. This results in a competitive strategy driven by a sequence of rapid maneuvers, or business actions, to adjust to changing conditions. To achieve this, we draw upon adaptive leadership theories from military science to argue that senior innovation executives must adopt an agile, iterative strategy that we term maneuver-driven competition. Maneuver-driven competition is driven by a never-ending "SROM" planning cycle of:

- *Scouting* the market landscape as well as the firm's current competitive position and capabilities.
- *Roadmapping* a set of appropriate innovation projects that can adjust the firm's products and capabilities to different scenarios as they unfold.
- *Orchestrating* the actions of subordinate innovation leaders by adjusting their goals and supplying the reasons behind these adjustments.
- *Maneuvering* the innovation firm by empowering subordinate innovation leaders and innovation workers to lead their projects toward these objectives without micromanagement by senior innovation executives.

The faster an organization can execute the SROM cycle, the better it can cope with the evolution of the innovation system, particularly when coping with the innovation butterfly. Executing the SROM cycle rapidly requires that all levels of leadership *modularize* the risks inherent in their tasks (or projects) so that they have sufficient latitude to respond to potential butterfly effects in an agile, flexible manner without disrupting other leaders' plans. Modularization of risk through timebuffers, cross-trained employees, knowledge management systems, and overstaffing, among other techniques, can reduce the probability that the failure or delay of one project can create a "domino effect" that derails the entire roadmap of projects. A final way to improve the speed and flexibility of the SROM cycle is for innovation firm's staffs to create a shared set of standardized procedures and information systems. Standardization has three benefits. It improves the ability of senior innovation leaders to predict how a given scenario will impact a project's outcomes. It also reduces the amount of time spent by innovation workers in "reinventing the wheel." Finally, it promotes an environment in which capabilities can be more rapidly recombined, because the innovation firm's personnel, which are what firms' capabilities are ultimately embedded in, can be more rapidly redeployed in a "plug and

---

[2] The Collected Wisdom of HERACLITUS, Translated by Brooks Haxton, Viking, New York, 2001 ISBN 0-670-89195-9.

play" manner as circumstances warrant. Combining the three tools of SROM-driven agility, modularizing risk, and standardization of processes provides the tool-kit with which the innovation firm can profitably navigate the ever-shifting innovation landscape and, even shape it to its own advantage.

Even using these tools, however, the risk of misalignment in terms of individual biases and goals resulting from the principle of providential behavior remains. Hence, we discuss in the last three chapters the work of the innovation leaders operating at all levels within the innovation firm: we compare their role with an architect, the captain of a ship, and the coach of an athletic team. Many of the examples cited in these chapters reinforce the lessons about nature of complexity in innovation systems and the tools necessary to cope with and take advantage of the innovation butterfly's effects. However, these chapters also point to the need for firm leaders to personally lead the development of a firm culture among innovation workers and management to support maneuver-driven competition. This culture has several characteristics. One is that it must support and reward employee personnel that act for the good of the firm as a whole while executing their project. Otherwise, the dark side of the principle of providential behavior—opportunism, misalignment, and other "bad behavior" detrimental to the firm—will become rampant. A necessary concomitant for empowerment is veracity in communication and a willingness to learn. Innovation leaders bear a special responsibility here, because "killing the bearer of bad news" just once can destroy the ability of the decentralized firm to meaningfully communicate and thus frustrate any attempts by senior leadership to orchestrate coherent action of individual innovation teams. In addition, innovation leaders must also ground their culture in a "love for problem solving," because this is what drives innovation workers to their highest levels of creativity. For example, awards in the form of raises and promotions, which drive most employees in other fields, are not nearly as effective with innovation workers. Instead, recognition for their technical capabilities and the ability to work on challenging innovation projects work better. Finally, the innovation firm requires a deep culture of empowerment of all innovation workers in order to make the most of the decentralization inherent in maneuver-driven competition.

Only when all these pillars—understanding, tools, culture, and leadership —are in alignment can the innovation firm respond in a sufficiently *agile* manner to maneuver and cope with the complex innovation system. In fact, once these are in place, the firm need no longer fear the innovation butterfly, but can in many cases harness its effects to competitive advantage.

Having said this, we realize that we have primarily leveraged the fields of software and discrete product innovation (e.g., cars, medical devices, video games) within this book. These fields, generally speaking, have fairly fast product lifecycles. Analogously, we believe that the bulk of the prescriptions in this book will apply to fast cycle services sector (such as professional consulting, financial services, or entertainment) as well. For those settings that have relatively slower lifecycles, such as education, transport, and most nonprofits, however, the concepts in this book may need significant modification. Applying these ideas to the level of macroeconomics also brings special considerations into the mix. Similarly, applying these ideas to start-up, especially venture capital backed firms in their earliest years, or family owned firms, may need special considerations such as how the principal's interests influences decision making.[3] On the other hand, the influence of the "whims" of founders at startups, much like senior managers at large firms, shapes the perception among many teams on what the organization will value. The three principles outlined in Part I can provide a starting point for managing innovation in these fields, particularly for services such as health care and public sector utilities and startup firms in the energy sector, all of which are well known to be badly in need of innovation.

Another important point that we do not address here is how and when to teach these concepts. Currently, they are primarily taught by mentoring, which perhaps explains why innovative industries tend to geographically cluster in areas such as Singapore, Silicon Valley, or Bangalore. Teaching innovation leadership in the classroom is another matter. Generally in engineering, computer science, or other schools that train innovation workers, only the basic scientific principles and particular types of technical skills such as computer-aided design (CAD) are taught as formal courses. In other words, it is as if we were teaching our innovation workers the blocking and tackling aspects of football without teaching them how to develop a game plan and adjust it, much less how to coach other innovation workers. Business schools attempt to teach innovation leadership to some extent with their entrepreneurship, human capital, strategy, and management of technology classes, but the individual aspects taught in these classes are never really integrated. And even if the classes were integrated, how would they capture the true extent of the complexity of the innovation system in a classroom environment?

The answer to the question of how to teach innovation leadership requires a great deal of thought. The complexity underlying innovation challenges is likely to multiply. In the absence of systemic thinking about the underlying science, evolving economic, and technical challenges, and the leadership skill to integrate these ideas, we as a society may find ourselves unable to create a world which can continue to innovate itself toward a better future.

---

[3]Cable, D.M., Shane, S.: A prisoner's dilemma approach to entrepreneur-venture capitalist relationships. Acad. Manage. Rev. (1997).

Appendix A
Analytics (Tracking Task, Project, Pipeline, and Portfolio Risks)

We review some of the relevant literature and identify opportunities for collecting data, along with the analytics needed for visualizing and tracking risk under the following categories:

1. Interdependence and risk within a single project
2. Roadmapping, triggers and sequencing
3. Stage-wise risk within a R&D pipeline
4. Aggregate evolution of portfolio risk

A.1 Interdependence and Risk Within a Single Project

Figure A.1 demonstrates the mapping between the information structure and the execution strategy of interconnected product development tasks.[1] The information dependencies between development tasks constitute the structure of the development process.

In this context, development activities are classified into three types: dependent, interdependent, and coupled.[2] Two tasks are said to be dependent if one task depends

---

[1] This figure has been taken from
- Joglekar, N.R., Yassine, A.A.: Management of information technology driven product development processes. In: Boone, T., Ganeshan, R. (eds.) New Directions in Supply Chain and Technology Management. Amacom Press (2002).

For allied details, see
- Yassine, A., Chelst, K., Falkenburg, D.: A decision analytic framework for evaluating concurrent engineering. IEEE Trans. Eng. Manage. 46(2), 144–157 (1999).
- Joglekar, N.R., Yassine, A., Eppinger, S.D., Whitney, D.E.: Performance of coupled product development activities with a deadline. Manage. Sci. 47(12), 1605–1620 (2001).

[2] Eppinger, S.D., Whitney, D.E., Smith, R.P., Gebala, D.: A model-based method for organizing tasks in product development. Res. Eng. Des. 6(1), 1–13 (1994).

E.G. Anderson and N.R. Joglekar, *The Innovation Butterfly*,
Understanding Complex Systems, DOI 10.1007/978-1-4614-3131-2,
© NECSI Cambridge/Massachusetts 2012

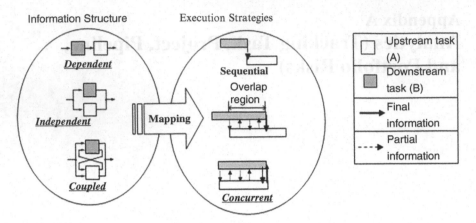

Fig. A.1 Information structure and execution strategies (From Joglekar and Yassine 2001)

on the other for input information. On the other hand, if both tasks depend on each other for input information, then these two tasks are coupled. Finally, if there is no information dependency between both tasks, then they are independent. The execution strategies employed in the development process determine the development process schedule. The information structure could be mapped to three different execution strategies via different rework (i.e., development iteration) risk levels. The sequential execution of development tasks requires that upstream tasks completely finish before downstream tasks can be started. In the overlapped execution strategy, upstream tasks are scheduled to start first but downstream tasks start before the completion of upstream tasks. Finally, the simultaneous start and finish of tasks characterize the concurrent execution strategy. There are several elegant models that analyze the cost, quality, and performance (and risks) associated with such strategies.[3]

Many tasks comprise of a project. When we scale up the analysis from tasks to an entire project, one powerful methodology in process management modeling that allows studying the flow of information among activities is the design structure matrix (DSM).[4]

In the DSM representation, the development process is modeled by a collection of interdependent development tasks. Each task receives information from other tasks, processes the information, and delivers information to subsequent

---

[3] For a review, see: Loch, C., Terwiesch, C.: Coordination and information exchange, Chapter 12. In: Loch, C., Kavadias, S. (eds.) The Handbook of New Product Development Management. Butterworth–Heinemann, Oxford (2008).

[4] Joglekar, N.R., A.A. Yassine.: Management of Information Technology Driven Product Development Processes, in New Directions in Supply Chain and Technology Management: Technology, Strategy, and Implementation. T. Boone and R. Ganeshan (eds.), Amacom Press (2002).

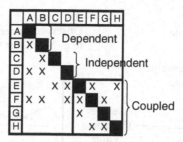

Fig. A.2 A representative design structure matrix (DSM) (adapted from Eppinger et al. 1994)

development tasks. A sample DSM is shown in Fig. A.2. The simplest DSM model of a design process is a binary, square matrix, where the "X" marks show task interdependencies. Lower diagonal elements in the DSM represent feed forward information flows among the tasks, and upper diagonal elements represent feedback. A feedback information flow captures the iteration and rework potential in a design process.

DSM representation readily reveals the existence of the three information dependency types, discussed in Fig. A.1, which allow for more efficient execution strategies. For example, the DSM in Fig. A.2 shows that tasks A and B are dependent since there is a dependency mark at the intersection of A's column and B's row, specifying this information dependency. Similarly, tasks C and D are shown to be independent since there is no mark at the intersection of C's column and D's row. Coupling between tasks E, F, G, and H is evident by the existence of a block indicating that each task in the block depends on the other tasks for information. Blocks in the DSM can be identified through partitioning (see www.dsmweb.org).

Sequencing algorithms allow for the reorganization of tasks in the matrix to provide an improved sequence. This new sequence increases the efficiency of the design process, reduces product development lead time, and allows for reduction in the project risk.[5]

The dependencies across complex innovations tasks result in iterative problem-solving cycles during innovation projects. The easiest way to understand the variation created by iterations is by tracking key design parameters, for instance, the fuel consumption measured in miles per gallon (MPG) during automotive design, as the problem-solving effort proceeds over time. Ideally, a design parameter would converge to a set target, thereby reducing the risk associated with the design as shown on the left-hand side of Fig. A.3.

---

[5] Browning, T.R., Eppinger, S.D.: Modeling impacts of process architecture on cost and schedule risk in product development. IEEE Trans. Eng. Manage. **49**(4), 428–442 (2002).

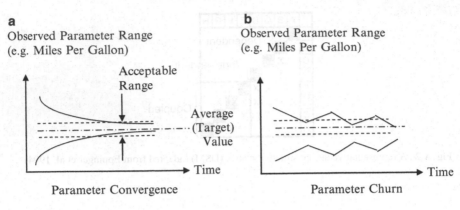

Fig. A.3 Parameter convergence and churn

However, such convergence is not guaranteed, and the situation can lead to an oscillating outcome, termed as churn, as shown on the right-hand side of this figure.[6] Such oscillations indicate unstable progress owing to planned or unplanned rework. This analysis assumes that the DSM is static, that is neither a new interconnection, nor a new task, is created. In reality, new uncertainly may be creeping in owing to a butterfly effect, for instance, the problem solving in a software development process may also creating new bugs. Conversely, none of the existing interdependencies are eliminated through iterations. An emerging trend in this setting is toward a dynamic analysis of DSMs, especially in distributed settings. An allied area that shows considerable promise for accounting for emergent phenomena is design for adaptability (DFA).[7]

An important aspect of tracking parameters in a complex design setting is to avoid a narrow specification of the target upfront in the process. Sobek et al. (1999) describe a method to model convergence based on observed-based practices in Toyota's product development process, called set-based concurrent engineering (SBCE). With SBCE, Toyota's designers think about sets of design alternatives, rather than pursuing one alternative iteratively. As the development process progresses, they gradually narrow the set until they come to a final solution.[8]

---

[6] In distributed innovation settings, where problem solving is carried out and synchronized periodically, churn cannot be avoided. See
- Yassine, A., Joglekar, N., Braha, D., Eppinger, S., Whitney, D.: Information hiding in product development: the design churn effect. Res. Eng. Des. **14**, 145–161 (2003).
- Mihm, J., Loch, C.H.: Spiraling out of Control: Problem-Solving Dynamics in Complex Distributed Engineering Projects. In: Braha, D., Minai, A., Bar-Yam, Y. (eds.) Complex Engineering Systems. Perseus Books, New York (2006).

[7] See, for example,
- Engel, A., Browning, T.R.: Designing systems for adaptability by means of architecture options. Sys. Eng. **11**(2), 125–146 (2008).

[8] Sobek, D.K. II, Ward, A.C., Liker, J.K.: Toyota's principles of set based concurrent engineering. Sloan Manage. Rev. **40**(2), 67–83 (1999).

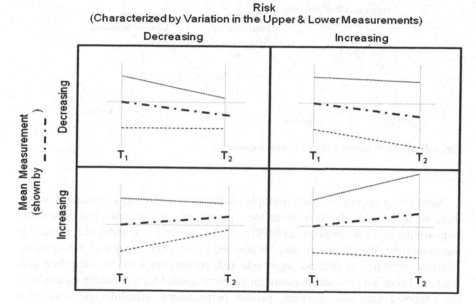

Fig. A.4 Alternate patterns of performance evolution over the time interval $T_1 T_2$

Since the teams are interested in reducing risk, it is worthwhile to track the mean value of a parameter along with its variance. We refer to the work of our colleague Professor Tyson Browning's methodology for tracking the evolution of key performance parameters (e.g., the MPG—for an automotive vehicle being developed) along with the project's projected cost and schedule.[9] According to Browning et al. (2002), the risk associated with a project can be computed by tracking the mean and the variance data. They term this as the risk value method (RVM), and also point out that measurements across time during development may indicate that the mean value of a parameter can rise, remain constant, or drop; and at the same time its variation may go up, remain constant, or reduce. Figure A.4 shows four of these patterns of evolution (this is a simplified version of a diagram analyzed by Browning et al.). The task of a team tracking the project risk is to understand the source of this shift in the mean (if any), along with the variation (if any). Sometimes it is difficult to track the source of this variation. Other times, the source can be attributed either to a planned experimentation strategy, or it unplanned decision, that might turn into a full blown butterfly effect.[10]

---

[9] For project level risk management during development, see:

- Browning, T., Deyst, J., Eppinger, S., Whitney, D.: Adding value in product development by creating information and reducing risk. IEEE Trans. Eng. Manage. **49**(4): 443–458 (2002).
- Higuera, R., Haimes, Y.: Software risk management. Software Engineering Institute Technical Report, CMU (1996).

[10] Thomke, S.: Experimentation Matters: Unlocking the Potential of New Technologies for Innovation. Harvard Business School Press (2003).

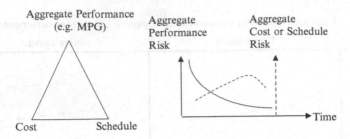

Fig. A.5 Tradeoffs between aggregate risk measures

Most complex projects track multiple attributes or parameters, especially when there are technical tradeoffs between these parameters, for example, it is difficult to improve the fuel consumption (in MPG) and the weight (in pounds) of a vehicle at the same time. Weight in turn may be affected by the overall payload. Given these multiple attributes of interest, aggregate risk parameters must be identified and tracked. There is a growing literature on the steps needed for translating variation in the observed data into aggregate project performance, schedule and cost risks (Browning et al. op cit). Indeed, certain customers, such as the Department of Defense, specify how the overall risks parameters are to be specified and tracked.[11]

Tracking aggregate risk of a project is as much an art as science. It involves performance, cost, and schedule tradeoffs as shown in the stylized Fig. A.5. For a deeper discussion of risk management in these settings, we refer the reader to Loch et al. (2006).[12] These authors argue for assigning individuals with the task of tracking and mitigating specific risks—such individual must understand that innovation projects can be ambiguous and uncertain. We augment their discussions by pointing out that in the presence of complexity, such individuals must also be charged with the recognition of early signs of the butterfly effect.

Typically, there are organizational gaps (and often geographic separation) between personnel (typically at a higher level) who are charged with tracking the aggregate performance and the bench level individuals (e.g., scientist, engineer, or a designer) who are conducting the detailed work. Oftentimes, bench level personnel are equipped with ability, or have access to data, to provide a causal explanation for the outcome. However, these individuals do not have the incentives to escalate their findings, especially when these findings are negative. A directed telescope, or a technical lead informing senior managers on the problems and status, as described in Part II, comes in handy in gaining visibility in these situations. This is important for the ongoing oversight for a project because enhancing complexity raises the

---

[11] Risk Management Guide for DoD Acquisition. (www.dau.mil/pubs/gdbks/risk_management.asp) (2006).

[12] Loch, C., DeMeyer, A., Pich, M.: Managing the Unknown: A New Approach to Managing High Uncertainty and Risk in Projects. Wiley (2006).

Resource Allocation Policy	Rework Probability (p =25%)		Rework Probability (p= 62.5%)	
	Duration (Days)	Reduction	Duration (Days)	Reduction
With Foresight	44.12	18.18%	89.12	8.94%
Directly Proportional	54.37		97.87	

Fig. A.6 Effect of Foresight (i.e. Not Ignoring Rework) on the Project Completion Time

probability of rework. For instance, the probability of rework is a not a fixed number across successive iterations. While it be ideal to minimize this probability through modular actions (e.g., during the design phase), it is not always possible to do so. For a given probability of rework, architects can achieve substantial reductions in the project schedule by using a foresighted (i.e., an optimal control policy, that admits the probability of rework by looking ahead), when compared with policies that allocates resources based on backlogs without accounting for the rework risk.[13] Figure A.6 shows that in simulation study, if the probability of rework was 25%, then effect of ignoring the rework, over an optimal policy that considered the rework, would be about 18%. However, if the probability of rework was 62.5%, all other things being equal, the effect would be about 9%.

A.2 Road Mapping, Triggers and Sequencing

In this section, we discuss how to plan and follow the evolution of risk as a firm goes through a sequence of projects. For instance, Fig. A.7 illustrates the roadmap for telematic products associated with an automotive firm. Such a map identifies platforms (e.g., to introduce a new platform in years 0 and 4). It also identifies hardware upgrade and software release cycles that might account for market gaps. Such maps are usually created by individual firms based on their perception of market need, competitive position, and technology capability. Platform planning and its relationship to modular architectures under uncertainty have received of lot of attention in the literature. However, the effects of the innovation butterflies in shaping the evolution of platforms is yet to see the same degree of research attention, in terms of architectural and market place evolution, and especially in terms of maneuver-driven competition.

A related document is a technology roadmap that is typically created by an industry association or at national laboratories. For instance, Fig. A.8 was assembled by the National Renewable Energy Lab to indicate the relative laboratory scale

---

[13] Joglekar, N.R., Ford, D.N.: Resource allocation policies with foresight structures and design concurrency. Eur. J. Oper. Res. **160**(1), 72–87 (2005).

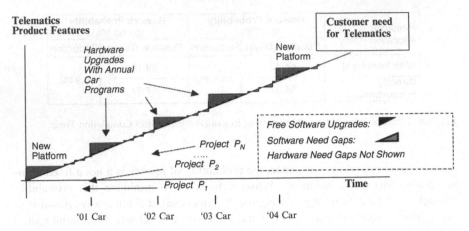

Fig. A.7 A telematics product road map at an automotive firm. From Joglekar and Rosenthal (2003), Coordination of design supply chains for bundling physical and software products. J. Prod. Innovat. Manage. **20**, 374–390

performance of photovoltaic solar panel technologies. However, both these types of roadmaps do not illustrate the evolution of risks explicitly.

Phaal et al. (2004) have argued for the creation of multidimensional maps that identify triggers and dependencies. We have modified these maps and incorporated technology and capability risks, buffers, and butterfly effects. For instance, left-hand panel of Fig. A.9 shows both the product and the technology maps. The right-hand side panel adds market drivers, capabilities needed to meet these drivers, technology developments, and the triggers, along with arrows that indicate dependencies. While the road mapping literature does not call out the need to superimpose risk-related data on to these maps, we have seen instances where firms require that risks (on the technical, market as well as HR—or capability—side) must be identified when these maps are presented during annual planning and portfolio review processes. Such maps provide opportunities for understanding of the evolution of the butterfly effects, and their consequences on the overall R&D program. In many instances, savvy managers insist on building time and resource buffers (or other risk hedging strategies) into these maps.

A.3   Stage-Wise Risk Within a R&D Pipeline

Figure A.10a shows an aggregate risk distribution by the size of the circle as it relates to market growth and margin. Figure A.10b shows a disaggregate view of these projects in three stages of development. Allocation of resources, the selection of targeted complexity, and screening mechanisms have received a lot of attention

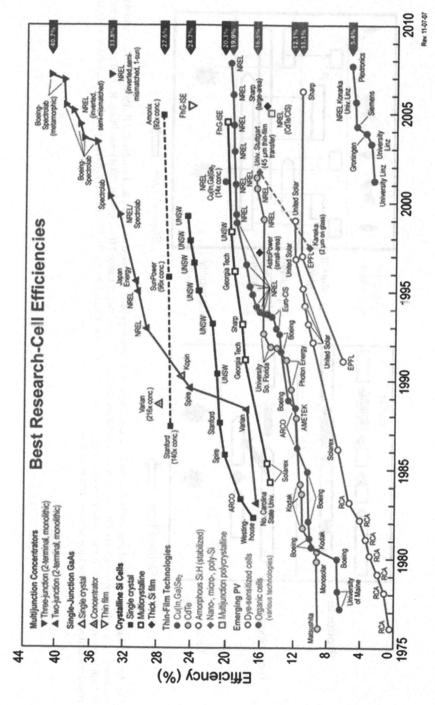

Fig. A.8 A roadmap: best obtained laboratory efficiencies for photovoltaic panels. Data compiled by Lawrence Kazmerski, National Renewable Energy Laboratory (NREL). The chart at the *right* illustrates the best laboratory efficiencies obtained for various materials and technologies; generally, this is done on very small, i.e., 1 cm², cells. Commercial efficiencies are significantly lower. Available at http://en.wikipedia.org/wiki/Solar_cell

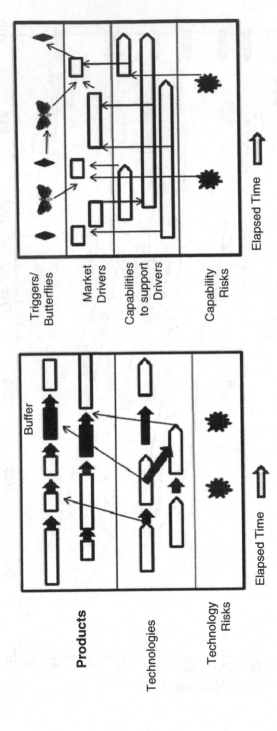

Fig. A.9 Multidimensional road maps with risks and triggers. These road mapping charts are based on Phaal, R., Farrukh, C.J.P., Probert, D.R.: Technology roadmapping—a planning framework for evolution and revolution. Technol. Forecast. Soc. Change **71**(1–2), 5–26 (2004); and Phaal, R., Muller, G.: An architectural framework for roadmapping: toward visual strategy. Technol. Forecast. Soc. Change **76**(1), 39–49 (2009). We have stylized the diagrams and superimposed butterflies on these time lines to show their emergence

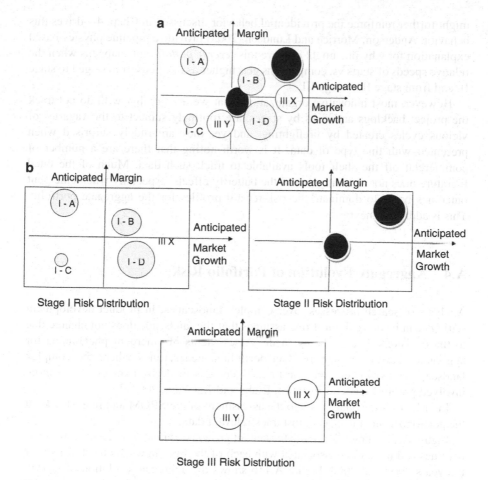

Fig. A.10 (a) Aggregate product portfolio by stages (size of the *circle* shows the amount of perceived risk. For a discussion of bubble charts, see Cooper, R.G., Edgett, S.J., Kleinschmidt, E.J.: Portfolio Management for New Products. Perseus, Cambridge, MA (2001)). (b) Disaggregate product portfolio by stages (size of the *circle* shows the amount of perceived risk)

in the literature (see Terwiesch and Ulrich 2009). It is worth noting that the portfolio risk, and associated backlogs of tasks, is supposed to reduce as one traverses from stage I to stage III. However, every so often, *reverse patterns* are observed—that is, large amounts of risks (and consequently backlogs) are accumulated at the back end (e.g., stage III in Fig. A.10b).

Repening et al. (2001) point out that this leads to firefighting that is allocating the much needed resources from the front end to the back end, that leads to persistent imbalance in the R&D pipeline. Repenning et al. also provide a behavioral argument for why such build-ups persist: in some settings, the hero mentality—which might further reinforce the providential behavior discussed in Chap. 4—drives this behavior. Anderson, Morrice and Lundeen (2005) provide a pipeline physics-based explanation for why one might observe this *reverse pattern*: this happens when the relative speeds of starts vs. completions gets higher as one goes from stage I to stage II, and from stage II to stage III.

However, most innovative organizations that we are familiar with do not track the project backlogs and risks by stages, are routinely subject to the vagaries of vicious cycles created by firefighting, and yet they are rudely surprised when presented with this type of data. It is worth noting that there are a number of commercial off the shelf tools available to track such data. Much of the cited literature does not explicitly model the butterfly effects, especially when emergent outcomes begin to dominate the risk-reward profiles for the aggregate portfolio. This is addressed next.

A.4   Aggregate Evolution of Portfolio Risk

Analysis of search processes, over complex landscapes, in product development settings can be carried out if one assumes that the landscape does not change due to the decisions that are being made. Assessments of emergent phenomena for aggregate analyses of new product portfolios, i.e., settings where the complex landscape evolves over time due to early choices made by managers, is a more involved problem (see, for instance, Braha and Bar Yam, 2007).[14]

In order to bring analytics into the assessment of the SROM and related risks at the portfolio scale, managers must track relevant data.

Figure A.11a shows the charts for market growth and market share for four product lines and the circles associated with each of the line shows the level of risk for the years 2007 and 2008. Figure A.11b shows the aggregate evolution of growth rate, market share, and portfolio risk. Aggregating the risk of an entire portfolio requires that an analyst exercise some judgment in how the risk in the individual bubbles is added to derive the overall performance. For instance, the larger bubble on the lower right-hand side of in Fig. A.11a may be a platform product, that deserves a higher weight than a niche (or a derivative) product.

Some organization, such as Ericsson, also track the portfolio of competencies (or skill types), in terms of headcounts as part to their planning process. Each bar in Fig. A.12 shows key competencies by skill type: software developers, system designers, network support, project managers, HR, etc. In some years, the firm may have a plan to reduce the overall headcount. In such a case, the organization can identify key bottlenecks. Any possibilities of some butterflies creating shifts in the demand for talent can be tested using such data.

---

[14] Braha, D., Bar-Yam, Y.: The statistical mechanics of complex product development: empirical and analytical results. Manag. Sci. **53**(7), 1127–1145 (2007).

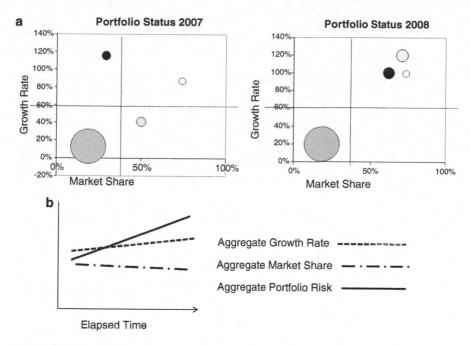

Fig. A.11 (**a**) Bubble charts with growth, market share, and risk (size of the *circle* shows the amount of perceived risk). (**b**) Aggregate evolution portfolio performance

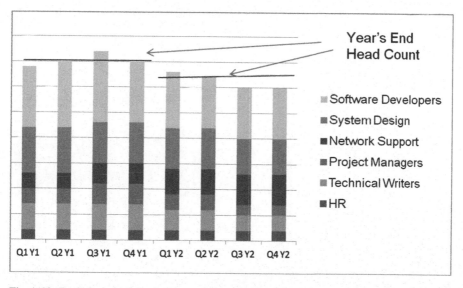

Fig. A.12 Evolution of required resources broken down by competence. This illustration is a stylized version, with altered scale, of a diagram by Miranda. Other diagrams that we have seen also show the projected vs. available resources, and along with internal and external nature of the resources. Miranda, E.: Strategic resource planning at Ericsson Research Canada. Presented to PMI Global Congress 2003 in the Hague, the Netherlands (2003)

Appendix B
Scenario Planning and Simulation

We have argued that it is difficult to forecast the outcome of innovation efforts with precision because of the potential for innovation butterflies. We have also identified distributed leadership, development of shared processes, and adaptive culture as key elements that enable organizations to shape and emergent outcomes. Scenario planning is a structured methodology for creating an understanding of and preparing for possible future states. Since the focus is on developing and understanding, rather than forecasting the probable outcomes, scenario planning can contribute to the development of distributed leadership, shared processes, and adaptive culture. In this section, we outline the steps associated with scenario planning processes and also discuss the opportunities of using simulation tools in this context. The roots of this methodology go back to Herbert Kahn, at Rand Corporation, who constructed structured forecasts with military applications in mind.[1]

Scenarios have been used extensively within projects to inform product development decisions, within large firms such as Shell to shape their strategies, and within an entire sector of the economy, such as the effort by the Intergovernmental Panel on Climate Change (IPCC) to determine the possible impacts of various assumptions about demographic, economic, and technical trends on the global environment.[2]

---

[1] For the history and managerial applications of scenario planning, see:
- Kahn, H., Wiener, A.J.: The year 2000: a framework for speculation on the next thirty-three years. Macmillan, New York (1967).
- Schoemaker, P., J.H.: Scenario planning: a tool for strategic thinking. Sloan Manage. Rev. **36**(2), 25–40 (1995).
- Lindgren, M., Bandhold, H.: Scenario planning. The Link Between Future and Strategy. Palgrave Macmillan, Hampshire (2003).

[2] For examples applications of scenario planning in product development, see:
- Noori, H., Chen, C.: Applying scenario-driven strategy to integrate environmental management and product design. Prod. Oper. Manage. **12**(3), 353-368 (2003).
- Robertson, D., Ulrich, K.: Planning for product platforms. Sloan Manage. Rev. Summer, 19–31 (1998).

E.G. Anderson and N.R. Joglekar, *The Innovation Butterfly*,
Understanding Complex Systems, DOI 10.1007/978-1-4614-3131-2,
© NECSI Cambridge/Massachusetts 2012

165

B.1   Scenario Planning Steps[3]

	Step	Activity
1	Define the problem	Define the conceptual and temporal boundaries for the construct.
2	Identify major stakeholders	Identify major stakeholders and actors who would have an interest in the issues under analysis.
3	Identify drivers of key factors	Make a list of current trends or predetermined elements that will affect the construct.
4	Identify key uncertainties	Identify key uncertainties whose resolution will significantly affect the variables of interest.
5	Describe future states for the scenario variables	Describe how scenarios variables might be in the future, project expected and most likely outcomes for them.
6	Construct initial scenario themes	One alternative is to construct two forced scenarios by placing all positive outcomes of key uncertainties in one scenario and all negative outcomes in the other.
7	Check for consistency and plausibility	Identify whether the combinations of trends and outcomes of the initial scenarios are indeed consistent and plausible. Eliminate combinations that are not credible or impossible, create new scenarios until you achieve internal consistency.
8	Assess and interpret the initial scenarios	Organize possible outcomes and trends around the scenarios. Make sure that the scenarios are strategically relevant and that they bracket a wide range of outcomes. Naming the scenarios is usually helpful.
9	Identify research needs	Assess the revised scenarios in terms of how the key stakeholders would behave in them.
10	Develop quantitative models	Reexamine internal consistencies of the initial scenarios and assess whether certain interactions should be formalized or investigated through quantitative modeling.
11	Evolve toward scenario analysis and planning	Iterate the above steps until you converge to scenarios that are relevant, consistent, archetypal, and (ideally) describe a future state which is somehow in equilibrium.

For a discussion of the scenario planning work at Shell:
- Wack, P.: Scenarios: uncharted waters ahead. Harv. Bus. Rev. **63**(5), 73–89 (1985).
- Wack, P.: Scenarios: shooting the rapids. How medium-term analysis illuminated the power of scenarios for Shell management. Harv. Bus. Rev. **63**(6), 139–150 (1985).
- Shell International Limited (2005). Shell Global Scenarios to 2025—the future business environment: trends, trade-offs and choices.
- For discussion of the scenario planning work associated with IPCC, see:
- Nakicenovic, N., Swart, R.: Emissions Scenarios. Cambridge University Press, London (2000).

[3] Many alternative templates for setting up a scenario planning process are available. Some describe a seven-step process, others describe variations. This table and allied discussion is based on a review article:
- Joglekar, N., Santiago, L.: Building theory and learning during scenario planning. A Boston University School of Management Working Paper (2010).

Scenario planning is a process that requires a fair amount of resources and thought. In some organizations, the process is owned within the central planning function, while others bring in external resources, typically consultants, to facilitate such processes. Since the goal for scenario planning is to prepare organizations for a "possible" future states, it has been seen as a key mechanism for fostering organizational learning. Arie De Geus and the thought leaders at the Society for Organizational Learning (www.solonline.org) have done a considerable work to develop the art and the science of scenario planning with many types of applications.[4] The extent to which scenario planning holds root and becomes a central part of innovation planning in terms of project, product, and portfolio management depends on organizational culture. Many large organizations have formal processes that embrace some parts of such thinking. We have seen other organizations where individuals or small groups engage in such learning in a systematic manner. On the other hand, it is rare to find larger teams, or entire businesses that are fully prepared for dealing with emergence in their planning processes.

B.2   Simulation: Group Model Building and Gaming[5]

Simulations tools can be deployed in a number of ways during the scenario planning. For example, simulation could be used to identify drivers, assess the impact of key uncertainties, describe future states, or to check of consistency and plausibility during the planning process. Large consulting firms routinely deploy simulation technologies to inform scenario planning. A related usage is the deployment of virtual environments, and gaming mechanisms, such that for stakeholders be they planners, architects, or coaches to develop shared mental models about the futures. There are two ways in which such mechanism can be productive. The first is to use the simulation tools as the basis for group model building exercises. One of the authors has deployed such exercises in settings such as a large insurance company, in which the marketing experts, planners, and product development experts came together and held discussions in terms of their own learning on how their respective decisions could interact and create butterflies. A detailed simulation could be built to capture these interactions and shared with their partners in the field who would sell these products.

---

[4] For a discussion of organizational learning at Shell, see
- De Geus, A.P.: Planning as learning. Harv. Bus. Rev. **66**(2), 70–74 (1988).
- De Geus, A.P.: The Living Company. Harvard Business School Press (2003).

[5] We refer the reader to texts that describe the use of simulations and gaming in order to facilitate organizational learning. For instance,
- Morecraft, J.D.W., Sterman, J.D. (eds.) Modeling for Learning Organizations. Productivity Press, Portland, OR (1994).
- Sterman, J.D.: Business dynamics: systems thinking and modeling for a complex world. McGraw-Hill/Irwin (2000).

In many instances, such exercises get terminated after identifying the model boundaries and relevant variables. In some instances, teams take such effort to the next level by building and calibrating simulation model. For instance, one of the authors has developed a system dynamics-based simulation models for examining various scenarios involving the development product, the placement of supply chain, and diffusion of sales of trucks in the Asian markets, on behalf of a large automotive firm. Such simulations are useful in accounting for risk with some precision. For instance, they can be used to size up the variability associated with the tipping point, while one is trying to assess the market share, and subsequent projected revenue for two alternative truck designs.[6]

---

[6] Anderson, E.G., Joglekar, N.R.: Managing Complexity in Distributed Innovation: A System Dynamics Perspective. System Dynamics Winter Camp, University of Texas at Austin (2007).

Appendix C
A Glossary of Complexity Terms

We define some of the terms that are needed to understand the key concepts in this book. Whenever it is difficult to provide succinct definitions, we describe related ideas, and provide simple examples to illustrate the terminology. An attempt has been made, while defining these terms, to remain consistent with standard terminology in the field of complexity science (see, for instance, Bar-Yam 1997).[1]

Activities: Actions that takes place within a system. These may consist of physical tasks, affect or thoughts that appear after observing a system or its components. An example of an action is a designer visualizing a color scheme for a room in a house using a computer-aided design (CAD) system.

Butterflies: Butterflies are atomic events or ideas that precipitate perturbations from a set plan. They are the cause associated with "the butterfly effect." We note that sometimes these perturbations are deliberate and mindful. For example, the architect for a residential building choosing an "A" frame construction, without a false ceiling, that creates a number of constraints for follow-on design choices (e.g., the height of the open ceiling, or the size of supporting columns), but also provides a low cost and open look to the space. Other times, that very decision (i.e., no false ceiling) could have an unintended, and perhaps undesirable, consequence: it may expose the electrical wiring in the line of sight of the inhabitants. Within the innovation system, these are called *innovation butterflies*.

Butterfly Effect: These are emergent formations. That is, they are typically created as unplanned side effect of small changes (a.k.a. butterflies) within a system. The magnitude of these effects could be small. For instance, the design of an appliance (e.g., washing machine) may add 1% cost premium, if it is offered with a special power saving feature. On the other hand, the magnitude of this effect could be large (e.g., "lead to a Tsunami in Texas"). For example, the power saving feature could

---

[1] Bar-Yam, Y.: Dynamics of Complex Systems. Perseus, Cambridge, MA (1997).

E.G. Anderson and N.R. Joglekar, *The Innovation Butterfly*,
Understanding Complex Systems, DOI 10.1007/978-1-4614-3131-2,
© NECSI Cambridge/Massachusetts 2012

dramatically amplify the growth of demand for eco-friendly appliances in the target market segments for washing machines. Within the innovation system, they are sometimes called *innovation butterfly effects* when appropriate.

Capability: Capability is the cumulative outcome of a set of activities. A known example of the evolution of capabilities in the automotive sector came about when firms hired and trained a cadre of electrical and software engineers to design electronic (i.e., microprocessor controlled) fuel injection systems (instead of mechanically controlled systems). This training created skills (know-how) and allowed the emergence of business processes to solve automotive problems using electronic technologies. These capabilities were subsequently used to create new features (such as antilocking control of brakes) using electronics and allied software algorithms that would not be possible using mechanically controlled device capability.

Complexity: A complex system contains a large number of mutually interacting elements. One way to think about and measure complexity is in terms of the information necessary to capture the macroscopic scale of interactions. A key part of this definition is the differentiation between microscopic scale and macroscopic scale. Another way to measure complexity is in terms of the uncertainty associated with the performance of the elements at the microscopic scale. Within a single development project, the elements at microscopic scale consist of tasks (such as design of the various subsystems in automotive development project). The overall project characteristics (schedule, time, aggregate resources for the entire development) are taken as the macroscopic view of the project.

Disruption: In the context of this book, an unplanned formation (q.v.) in the system that threatens to drive the innovation system into another regime of behavior. It is essentially synonymous with emergence (q.v.).

Distributed Innovation: The development of innovation projects by groups that are dispersed, either geographically or organizationally (or both). Common examples of distributed innovation include offshoring and outsourcing.

Element: Either a task or a component in at the scale of a single development project. Each single project is an element, when the scale is a portfolio of projects.

Emergence: Emergence is an unplanned formation (q.v.) within a system. For instance, in the electronic capability mentioned above (see the definition of capability) allowed the automotive firms to develop and implement antilocking breaks, and more recently incorporate software-based controls into the dashboard.

Escalation of Expectations (*Principle of*): When innovative outcomes (e.g., product performance) are built up as cumulative—and typically nonlinear—effects of effort expended in the past, the customers raise their expectations and continually demand improved performance. Under some special circumstances—known as disruptive innovations—the market mechanisms reset the performance metric and allied level of cumulative effect that are germane to a particular innovation system.

Exchange (*Principle of*): In distributed innovation systems, management solutions to local problems, even if they are effective, may result in a set of emergent outcomes elsewhere in the system.

Formation: This is an outcome (either a capability or a product) seen at the system level. It can either be emergent or be foreseeable.

Fractals: Patterns that repeat themselves at different scales.

Information: Relevant data that must be exchanged before a development task (or a project, at the portfolio scale) is carried out, or the outcome of such a task (or project).

Innovation: Solving a problem using the creative process. In product (or service markets), the quality of the solution is judged in terms of the reception that the product (or service) receives in the market place.

Innovation Butterflies: See "butterflies" (q.v.).

Innovation Butterfly Effect: The "butterfly effect" (q.v.) as it emerges within innovation systems.

Innovation Employee: Any person involved in innovation work from bench technician to marketing professional to chief technology officer.

Innovation Executive: A high-level innovation leader (q.v.) who manages a portfolio of innovation projects, each of which has its own subordinate innovation leader.

Innovation Leader: Often globally scattered individuals, be they bench scientists; engineers; technical leads in the field; architects; or project managers who worry about day-to-day coordination; product line planners and business strategists charged with the growth of a portfolio of products; directors of R&D, supply chain, or customer support; VPs of marketing, engineering, or human resources; or CEOs concerned about the survival and growth of their entire organization or its entire value chain.

Innovation System: Innovations are the result of entrenched processes. wherein competitive actions, employees' skill sets, interactions among various products in a portfolio, market needs, and investment policies are linked into a complex, multi-layered system, which we refer to for brevity as the innovation system.

Innovation Workers: Often globally scattered individuals, they are those who work on innovation projects and in whom are embedded the capabilities (q.v.) of a firm. Archetypical innovation workers are engineers, software programmers and architects, operations managers, and market researchers charged with helping define a new product, and their equivalents in the service field.

Interactions (*and their strength*): The (degree of) dependence between any two elements.

Maneuver-Driven Competition: Adaptive adjustment in project (and portfolio) strategies based on a Scout–Roadmap–Orchestrate–Maneuver (SROM) cycle in order to create and sustain competitive advantage.

Modularity: Specification of interfaces such that parts (or processes) across these interfaces can be developed in nearly independent manner.

Portfolio: A group of projects, or group of capabilities, that have a defined structure in terms of sequence of execution, and whose performance is planned/tracked in terms of variables such as return on investment (ROI), risk, etc.

Profile: A multidimensional measure of performance. The profile for a single R&D project may include time, cost, and quality measures.

Project: A set of tasks that have a defined structure in terms of sequence of execution, and whose outcome is planned and tracked in term of performance variables such as time, cost, and quality.

Providential Behavior (*Principle of*): Individuals (or groups) exhibit foresight, and biases, when managing their decisions related to complex innovation in distributed settings, based on their own perceptions and desires for the future.

Rework: The process of repeating a task, either because a new problem solving strategy needs to be tried, or because previous solution has an error.

Risk: Describes the degree of the deviation in one or more measures within a performance profile (for an innovation task, project or a portfolio) from the expected value. Typically, the goal of an innovation task (or project or portfolio) is to enhance the expected value while reducing associated risk. Emergent formations, such as the butterfly effect, typically increase the ambient level of risk.

Roadmap: A document that lays out the planned evolution of products and/or services and/or technologies and/or capabilities.

Scale: The complexity of the system depends on its unit of analysis (a.k.a. scale). In this book, we typically deal with three types of scales: tasks, projects, and portfolio. Of these three, tasks have the shortest time scale, and the portfolios have the longest time scale, associated with the evolution of underlying patterns. In general, in order to recognize patterns, teams must resort to different types of measures at different scales. For instance, a project may be tracked in terms of its cost, quality, and performance (e.g., technical or ROI) metric. A portfolio on the other hand, may be tracked in terms of overall ROI, market share, and growth metric. Based on the choice of metric, the complexity profile is monotonically decreasing the function of scale. That is, the information needed to describe a system at a larger scale must be a subset, through aggregation, of the information needed to describe that system on a smaller scale.

Scenario Planning: A planning methodology that focuses on a set of possible outcomes, rather than a set of probable outcomes. A central idea behind constructing such scenarios is to promote organizational learning, and thereby foster agility responses to emergent outcomes.

Socio-Technical System: A system involving the interaction of human and machine elements.

Time: The duration of relevance for an atomic task, project, or portfolio (also see scale).